W9-CKE-371

# SOLID WASTE MANAGEMENT AND THE ENVIRONMENT: THE MOUNTING GARBAGE AND TRASH CRISIS

Homer A. Neal
J.R. Schubel

STATE UNIVERSITY OF NEW YORK
AT STONY BROOK

Prentice-Hall, Inc. Englewood Cliffs, New Jersey 07632

Library of Congress Catalog Number: 86-63350

Editorial/production supervision: Barbara Marttine Webber
Cover Design: Marie Gladwish
Manufacturing buyer: Rhett Conklin



Printed in the United States of America

10 9 8 7 6 5 4 3 2 1

ISBN 0-13-822891-4 025

Prentice-Hall International (UK) Limited, *London*
Prentice-Hall of Australia Pty. Limited, *Sydney*
Prentice-Hall Canada Inc., *Toronto*
Prentice-Hall Hispanoamericana, S.A., *Mexico*
Prentice-Hall of India Private Limited, *New Delhi*
Prentice-Hall of Japan, Inc., *Tokyo*
Prentice-Hall of Southeast Asia Pte. Ltd., *Singapore*
Editora Prentice-Hall do Brasil, Ltda., *Rio de Janeiro*

## DEDICATION

The authors dedicate this work to

Jeannie, Margaret and Sarah Cynthia Sylvia Stout,

two of whom would not take the garbage out.

---

*SARAH CYNTHIA SYLVIA STOUT*

*WOULD NOT TAKE THE GARBAGE OUT**

*Sarah Cynthia Sylvia Stout*
*Would not take the garbage out!*
*She'd scour the pots and scrape the pans,*
*Candy the yams and spice the hams,*
*And though her daddy would scream and shout,*
*She simply would not take the garbage out.*
*And so it piled up to the ceilings:*
*Coffee grounds, potato peelings,*
*Brown bananas, rotten peas,*
*Chunks of sour cottage cheese.*
*It filled the can, it covered the floor,*
*It cracked the window and blocked the door*
*With bacon rinds and chicken bones,*
*Drippy ends of ice cream cones,*
*Prune pits, peach pits, orange peel,*
*Gloppy glumps of cold oatmeal,*
*Pizza crusts and withered greens,*
*Soggy beans and tangerines,*
*Crusts of black burned buttered toast,*
*Gristly bits of beefy roasts...*
*The garbage rolled on down the hall,*
*It raised the roof, it broke the wall...*
*Greasy napkins, cookie crumbs,*
*Globs of gooey bubble gum,*
*Cellophane from green baloney,*
*Rubbery blubbery macaroni,*
*Peanut butter, caked and dry,*
*Curdled milk and crusts of pie,*
*Moldy melons, dried-up mustard,*
*Eggshells mixed with lemon custard,*
*Cold french fries and rancid meat,*
*Yellow lumps of Cream of Wheat.*
*At last the garbage reached so high*
*That finally it touched the sky.*
*And all the neighbors moved away,*
*And none of her friends would come to play.*
*And finally Sarah Cynthia Stout said,*
*"OK, I'll take the garbage out!"*
*But then, of course, it was too late...*
*The garbage reached across the state,*
*From New York to the Golden Gate.*
*And there, in the garbage she did hate,*
*Poor Sarah met an awful fate,*
*That I cannot right now relate*
*Because the hour is much too late.*
*But children, remember Sarah Stout*
*And always take the garbage out!*

---

* From "Where the Sidewalk Ends", by S. Silverstein. Reprinted with permission of Harper, Row and Co.

# CONTENTS

# Figures

# Tables

# PREFACE

Urban population growth, increased per capita solid waste generation, increased permanence of chemicals and decreasing space for landfilling combine to set the stage for a major change in the way metropolitan areas must deal with solid waste disposal. Economic and environmental concerns highlight the question of what accommodations should be made to protect groundwater sources, surface water supplies and the atmosphere while sustaining the desired lifestyle of modern society.

In the United State more than 150 million tons of solid waste are produced each year. Most of this waste is disposed of in landfills that are not properly maintained, causing the contamination of both surface water and groundwater supplies required for human use. Some areas, such as Long Island, New York, are under direct legislative mandates to close all landfills within the next few years. This leaves only exportation, mass incineration, and recycling for waste disposal and each of these has its own complex technical and public policy problems. This monograph provides a brief overview of the considerations that must be taken into account in assessing the options that exist for solid waste disposal, and suggests a series of research initiatives to provide the basis for future decisions on this continuously expanding problem.

Much of the information on which this monograph is based was developed by a team of faculty and students at the State University of New York at Stony Brook Marine Sciences Research Center in a survey project conducted during the summer of 1985 and funded by the Ogden Corporation. This project resulted in six working papers reviewing the various disposal options and the historical aspects of solid waste generation. These papers are available upon request.

This text is prepared in conjunction with the initiation of a Waste Management Institute at the State University of New York at Stony Brook. The Institute will continue the studies conducted over nearly two decades by the Marine Sciences Research Center at the University to examine various aspects of waste disposal, including the first use of fly ash from incineration processes to manufacture blocks suitable for construction of artificial fishing reefs.

We are deeply grateful to the students and faculty who participated in the Working Paper Project: James Mackin, Myrna Jacobson, Bernice Malione, Dawn Rivara, Kathryn Schubel and Kenneth Swider. We also thank Avan Antia, Joanne Arenwald, Louis Chiarella, Norman Itzkowitz, Gregory Marshall and Miguel Olaizola for tracking down references and checking data. The numerous drafts were typed by Lisa Mayer and Gina Anzalone and the figures were prepared by Marie Eisel and Marie Gladwish. We gratefully acknowledge the copy-editing work of Jenifer Slater. We also thank Homer A. Neal, Jr. for the computer programming required in preparing the manuscript in camera-ready form. Special acknowledgements are due Dr. Blair Kinsman and Dr. Frederick Seitz for their critical review of the entire monograph.

H. Neal, J. Schubel
Stony Brook, New York
September, 1986

# FOREWORD

I first became acquainted with Stony Brook's Marine Sciences Research Center (MSRC) and Dean Schubel and Provost Neal three years ago when I was invited to serve on the Center's Visiting Committee. This is a distinguished and distinctive oceanographic institution. It focuses its attention on the coastal ocean and on the use of science to develop strategies to conserve and, when necessary, to rehabilitate coastal environments throughout the world. Scientists in the Center have devised some of the most innovative and effective strategies I have ever been exposed to.

Recently, I had a hand in creating a Waste Management Institute within the MSRC to institutionalize and expand the outstanding and creative work they have done for nearly two decades on waste disposal problems. Members of the Center's staff have turned what could have been serious environmental problems into creative opportunities to enhance the environment. They have built artificial fishing reefs out of stabilized coal wastes and now are testing strategies to stabilize the ash produced by burning garbage and trash. They have developed and tested management plans to combine submarine sand mining with disposal of contaminated materials in the sub-seabed pits, followed by capping with clean material to restore the seafloor to its original configuration. They have worked with every waste product from sewage to radioactive wastes. Their new institute is worth watching. Neal and Schubel's new book is worth reading.

I recall a Peanuts comic strip of a number of years ago that said there is no problem so large and complicated that it can't be run away from. The garbage and trash problem is an exception. This problem has become so large and so complicated that it can no longer be ran away from. It must be confronted head on and immediately, if we are to avoid a crisis. Many urban areas of the country, particularly along our coasts, are rapidly running out of landfill space and have no alternative disposal strategy in its place. It is a national disgrace that so little has been done to avert this crisis. In most parts of the country we deal with our municipal solid waste the same way our ancestors did when this country was first settled. We throw it into a dump. We may cover garbage and trash with a thin layer of dirt every day, but little else has changed in our disposal strategies. The per capita production and the nature of our wastes, however, have changed dramatically. Today, every man, woman, and child produces an average of nearly two-thirds of a ton of junk each year. This material is loaded with exotic compounds we created through modern technology. Many have long lifetimes and uncertain degradation products, which come from products we buy which wear out too soon.

I am reminded of Willy Loman's comment in the play "Death of a Salesman": "Once in my life I would like to own something outright before it's broken! I'm always in a race with the junkyard! I just finish paying for the car, and it's on its last legs. The refrigerator consumes belts like a goddam maniac. They time those things. They time them so when you've finally paid for them, they're used up."

We not only make things that don't last as long as they should, we don't recyclye them when we are finished with them. By failing to recover valuable materials through recycling, we waste energy and squander precious natural resources. We also add to the large and growing environmental burdens of waste disposal and of pollution associated with the mining and processing of virgin materials. Then there is our preoccupation with packaging. No other nation in the world spends as much on packaging as does the United States. The costs are passed on directly to the consumer. They also add to the staggering costs of waste disposal, costs which in many areas are second only to education. According to Neal and Schubel, packaging accounts for a very significant fraction of the total municipal solid waste stream.

We must be assidious in our search for solutions to the nation's garbage and trash dilemma. Neal and Schubel have shown us our choices. We must insist that our decisionmakers use the best information we have at any time to select the best strategy or combination of strategies to deal with the problem. Management is a dynamic, interactive process. As new information is developed, it should be folded in to the decisonmaking process to ensure that at any given time the best choices are made. Garbage and trash don't wait. More is coming all the time and there will be no end to it. An investment in new knowledge to deal with these problems more effectively would pay large dividends to society -- in reduced costs in a better and cleaner environment, in public health. A properly designed and coordinated research effort with partnership by the federal government and industry would seem to be appropriate. All segments of society stand to benefit from increased knowledge and improved technology. A concerted effort could enhance the chances of important advances and at greatly reduced cost.

In our search for solutions we must not overlook the obvious. We must do everything possible to reduce the magnitude of the problem through source reduction and recycling. The remainder should be managed with strategies which make full use of the knowledge we have and of the best technologies at hand. The authors point out that the best solution for Nashville may not be the best solution for New York City. In every case, however, the solution selected must conserve our environment and protect human health at an acceptable cost to the economy. Taxes which encourage the use of virgin materials and which discourage the use of recycled materials need to be changed. Those who produce the wastes, and the secondary producers, the consumers, should bear the costs of their dispsoal.

As with any problem, the nation's municipal solid waste problem must be recognized as a problem, and must then be stated in tractable form before it can be resolved. The authors have demonstrated vividly the magnitude and complexity of the problem and have outlined the boundary conditions which must be satisfied by any solution. They also have taken the next step in creative problem solving: They have outlined the alternative ways of attacking the problem and have provided an unemotional and objective assessment of the advantages and disadvantages of each alternative. This book will be of interest to all those interested in garbage and trash and should be required reading for anyone who has anything to do with setting and implementing policies for municipal solid waste management.

William E. Simon
September, 1986

---

*William E. Simon is Chairman of the Board of the Wesray Corporation and of the Board of Trustees of the U.S. Olympic Foundation. Mr. Simon was Secretary of the Treasury from 1974 to 1977, Chairman of the Economic Policy Board, the Federal Energy Office and the East-West Trade Board. Mr. Simon has received numerous honorary degrees and awards for his extensive public service.*

Fig. 1-1 Garbage accumulation during a strike of workers in New York City.

# CHAPTER 1

## A VIEW FROM THE TOP OF THE MOUNTAIN

How should society dispose of the huge mountains of garbage and trash it creates each day? This question encompasses some of the most complex and urgent environmental issues of our time. Policies and practices associated with the disposal of solid wastes have been important issues since humans first formed settlements. Waste products are a natural and inevitable result of living, but the volume of wastes and their composition, as well as the options for their disposal, depend upon population, lifestyle, and technology, and these have changed dramatically over time, particularly in industrialized nations. The industrialized world generates ever increasing quantities of solid waste, with a growing proportion comprised of complex synthetic chemical compounds whose disposal creates extraordinary technical, health, environmental, economic and political problems.

We are here concerned with garbage and trash. Because these waste products create particularly serious problems for municipalities, they are often referred to as municipal solid wastes. For brevity, they are also frequently simply called solid wastes. These are wastes from homes, offices, stores and schools. This category does not include wastes generated by industry.

### 1-1 GARBAGE AND TRASH PRODUCTION IN THE UNITED STATES

The United States produces more garbage and trash than any other nation on earth. And the amount is increasing. In 1980, the United States produced 140 million tons of solid waste.[1] This means that the average individual generated nearly 1200 pounds of solid waste that year. By the beginning of 1986 the United States was producing garbage and trash at a rate of nearly 150 million tons per year. An increasing fraction of this waste is made up of components which are non-biodegradable and, therefore, if not actively destroyed, represents an environmental burden which will persist for many future generations.

If the roughly 150 million tons of solid waste generated in 1985 were compacted in garbage trucks the resulting volume would be about 600 million cubic yards.[2] If spread uniformly over the land area of the State of Rhode Island it would form a deposit 0.5 feet thick. At today's rate of solid waste generation, if Rhode Island were the nation's landfill and if there were no further compaction or degradation of the garbage, that state would increase in elevation by more than 50 feet per century.

The volume of waste produced by the United States has increased dramatically over the past three decades and, according to projections, will continue to increase. We are vividly reminded of just how much waste we generate daily when, for whatever reason, the waste collection for a municipality is suspended; Fig. 1-1 portrays the accumulation of garbage and trash in New York City during a strike of its garbage workers.

Between 1965 and 1995 the population of the United States is forecast to grow by 35 percent, from 193 million to roughly 260 million.[3] If per capita production of solid waste remained unchanged, its total volume would increase by the same percentage. Projections indicate, however, that the generation of municipal solid wastes will increase even more rapidly than population.[4,5] (See Fig. 1 - 2)

We must get rid of this enormous mass of garbage and trash, but is it possible to diminish the severity of the problem through source reduction or recycling?

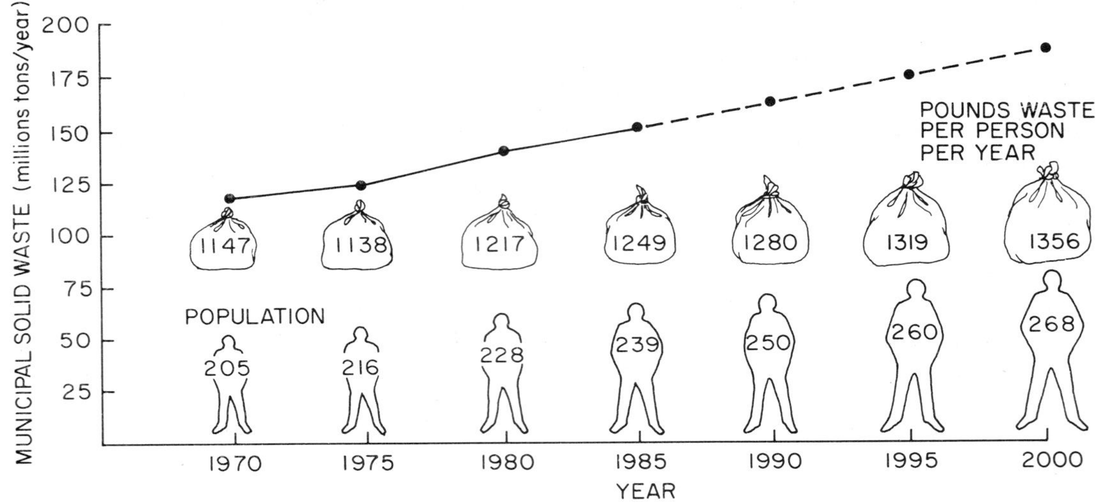


Figure 1-2 Municipal solid waste generation in the United States.

## 1-2 RECYCLING, SOURCE REDUCTION AND DISPOSAL

At present we recycle, or burn to generate heat or electricity, only about 8 percent of all the solid waste we collect.[6] This leaves 92 percent for disposal. Even the recycling rate for metal is low. About 70 percent of all metal is used just once and then discarded; only 30 percent is recycled.[7] Max Spendlove, research director of the United States Bureau of Mines, has observed that with all the cans, coat hangers, bottle caps, pots and pans and other metal objects we discard "... our refuse is richer than some of our natural ores."[8]

Clearly, any rational strategy for solid waste management must take into account opportunities for source reduction and recycling, as well as technologies for disposal. The advantages and challenges of recycling and source reduction are discussed in Chapters 2 and 8.

## 1-3 DISPOSAL OPTIONS

The primary solid waste disposal options are landfilling (Chapter 4), ocean dumping (Chapter 5), and incineration (Chapter 6). Each has come under increased scrutiny as changing regulatory laws, new technical findings, growing community concerns and needs, and dwindling available space have created mounting pressures on local, regional and national leaders to identify long term solutions to our solid waste problems.

### Landfilling

Landfilling is an extension of the age-old practice of hauling garbage to a dump, a site reserved for refuse at the edge of a community. Since antiquity this approach accomplished the goal of separating the community from the wastes which it no longer wanted to have near it. In the best of circumstances, the mounds of garbage decomposed at a rate that permitted a relatively small area to accept the refuse generated by a community for an extended period of time. While in use, dumps appear to affect only the specific areas in their immediate vicinity. But, as a result of wind and erosion, the impacts with time have been significant over broad areas, even in ancient cities when most of the wastes were biodegradable. Archaeological records reveal that cities such as Rome have risen in average elevation by tens of feet over the centuries as a result of one generation building additions to the city on top of garbage deposited by previous generations.

In comparison with garbage dumps of today, ancient dumps contained simple components. All items were either naturally occurring or derived from the natural environment without significant chemical alteration. Nevertheless, such sites usually were odorous, unsightly and unhealthy. Vermin that visited the dumps often frequented nearby households, transmitting diseases and plagues over large areas. Odors from decaying organic matter created unpleasant conditions even for persons living miles from a dump.

During the Industrial Revolution, from 1700 to the early 1800s, garbage disposal took on added significance. Before that period most people lived and worked in rural areas. But the invention of the steam engine, the spinning machine, and the weaving machine, and the development of the textile and steel industries set the stage for large numbers of people to crowd together in cities where job opportunities were more plentiful. This rapid build-up of population in cities led to widespread pollution of the soil, air and streams in most industrial areas because of the inadequacy of existing systems for the collection and disposal of the massive amounts of garbage and trash, as well as the human and industrial wastes generated by the large populations.

Over the past fifty years, there has been considerable improvement in the practice of landfilling of garbage, primarily because of the advent of large earth-moving machines which make it possible for garbage to be promptly covered with soil. Such landfills, commonly referred to as sanitary landfills, significantly reduce odors and access by vermin, and enhance decomposition of degradable material. But other deleterious effects remain, including the potential for contamination of surface and groundwaters by leaching of material from the landfill. Although engineering techniques exist for minimizing such problems, few landfills employ them. For example, in New York State, fewer than 20 percent of approximately 500 landfills have, or could qualify for, the proper environmental permits.[9]

Landfilling continues to be the most common disposal method nation-wide, but in many urban areas existing landfills are nearing capacity. Acceptable new sites are difficult to identify and often nearly impossible to secure.

### Ocean Disposal

Cities near the ocean have, at least in principle, another option -- to dump their refuse into the sea. Although at present there is no ocean dumping of municipal solid waste by the United States, there was in the past. And a number of regions continue to dump other wastes into the ocean. Each year the City of New York transports by barge and dumps an average of about 7 million tons of sewage sludge and 8 to 10 million cubic yards of dredged material offshore in the New York Bight.[10] The dreged material is dumped at a site 12 miles offshore. Through 1985 the sewage sludge was dumped at an adjacent site. New regulations require the relocation of the sewage dump site to a new site 106 miles from shore. An immediate question is what is the ability of the ocean -- or at least segments of it -- to accept large quantities of waste on a continuing basis without suffering long-term detrimental impacts on the coastal environment, on marine life, and eventually on human health.

### Incineration

Incineration as an option for municipal solid waste disposal also has its origins in antiquity. Indeed, one method used to increase the lifetimes of ancient dumps was to reduce the volume of waste by setting dumps afire. This practice, which often led to fires that burned for weeks, was utilized until quite recently (the 1950s). The stench created by burning, and the plume of polluted air which frequently extended back to the community from which the garbage came, led to regulations which eliminated burning in open dumps and in simple incinerators in most communities.

The modern version of the incineration process involves controlled burning at very high-temperature (more than 1000 degrees Fahrenheit) in special facilities called resource recovery plants. These facilities create essentially no visible emissions and no odors, and produce ash as the only solid residue requiring disposal. Moreover, some of the energy produced by burning the trash can be recovered and used to produce steam for heat or electricity, and various metals and glass can be recovered for recycling. But even this alternative is not without uncertainties. Concerns about high temperature incineration focus on the detailed chemical makeup of the gaseous emissions, their possible health hazards, and on methods for safe disposal or utilization of the residual ash.

## 1-4 THE WIDER ENVIRONMENTAL CONTEXT

Of the approximately 150 million tons of refuse generated in the United States in 1985, between 5 and 10 percent was recycled. Of the rest, approximately 80 to 90 percent (by weight) was disposed of in landfills; and 7-10 percent was incinerated. The remainder ended up in open dumps and undesignated sites.[11]

The options for disposal of municipal wastes are limited in variety and in number. Moreover, in some areas, the number of options is being reduced even further. To make matters worse, solid waste generation rates are increasing, and the composition of the waste stream is such that less and less of the material is biodegradable. Clearly, issues surrounding solid waste management should be regarded as critical and should command increased attention for coordinated research and planning and public education. The generation of solid wastes will continue, and society needs to take a longer, more creative view in its search for solutions.

Many cities are rapidly approaching a garbage and trash crisis: no place to put it. In 1934 New York City had approximately 17,000 residential incinerators, 22 municipal incinerators, and 89 landfills to handle its solid wastes. In 1985 it had no residential incinerators, three municipal incinerators and only three landfills. One of its three landfills (Fountain Avenue) closed at the end of 1985. It was full, after reaching a height of 145 feet. The Edgemere landfill receives only about 500-700 tons per day of the 27,000 tons per day generated by New York City. The other (Fresh Kills on Staten Island)--the largest landfill in the world with an area of more than 3,000 acres--has a "life expectancy" of about 16 years at the current rate of filling. At that time it will be 500 feet high. In only 16 years, landfill space in New York City will, for all practical purposes, be exhausted.[12] The New York City situation is not unique, or even unusual, among large coastal cities in the United States. Seattle, Washington, for example, is expected to run out of landfill sites within 3 to 5 years.

On Long Island (New York) availability of space is not the primary problem; contamination of groundwater is. Long Island's aquifer system is the sole source of drinking water for all of the more than 3 million residents of its Nassau and Suffolk Counties. Because of recurrent problems of groundwater contamination, the State enacted legislation in 1983 to stop by 1990 all landfilling of garbage and trash on Long Island and to permit landfilling of ash from resource recovery facilities only at selected sites.

To evaluate the present and projected impacts of solid waste disposal, one must focus on the extent to which each disposal method affects the public health, the purity of the atmosphere, the surface water, the groundwater, and the oceans. Aesthetic and economic factors must also be considered. There are advantages and disadvantages associated with each solid waste management option, and these differ from one location to the next. Each has impacts on our environment. The earth has a finite capacity to support human life. That capacity is determined in large measure by the purity of our air and water.

### The Atmosphere

The 20-mile thick layer of gases which encompasses the earth constitutes our atmosphere. It consists of approximately 78 percent nitrogen, 21 percent oxygen, and small amounts of carbon dioxide, water vapor, and other constituents. Other gases or particulates injected into the atmosphere represent forms of pollution whether they are from natural or anthropogenic sources. As a result of the great industrialization that has occurred worldwide in the past century and a half, humans have degraded air quality in some local areas enough that human health is threatened, and we have changed the composition of the atmosphere on a worldwide basis to the extent that it may alter the earth's climate.

Each primary chemical element and compound in the atmosphere plays a key role in supporting life. Carbon dioxide is required for photosynthesis and therefore is essential to life processes. Oxygen is the fuel of life. The chemical oxidation process that occurs in animal cells, for example, provides the energy that sustains the metabolic processes. The highly reactive nature of oxygen, as manifested in the rusting of iron and in the rapid progression of combustion in open air, is vital to the processes of life. Nitrogen plays a critical role in the formation of protoplasm, the cellular substance found in all living things. The blend of these atmospheric components determines the delicate balance of life on earth.

In the urban setting, hundreds of compounds and elements other than oxygen, nitrogen, carbon dioxide and water vapor are continually released into the atmosphere. Many of them individually may produce undesirable effects on biological systems, including humans. Under certain conditions, two or more of the emissions may react and produce even more noxious components. And the concentrations do not have to be large for the biological consequences to be significant. At levels of less than one part in a billion parts of air, certain pollutants can cause serious public health hazards.

In Chapters 3, 4, and 6 we examine the extent to which pollution from solid waste disposal may influence the overall quality of the air we breathe.

### The Hydrosphere

A cube of roughly seven hundred miles on a side encompasses a volume roughly equal to the total volume of water on and in the earth. However, not all of the earth's water is directly useful for primary human activities.

Over 97 percent of the earth's water is contained in the world ocean which blankets more than 70 percent of the earth's surface. This water is not directly useful for drinking or for many other human purposes, but it is the ultimate source of our drinking water because evaporation from the ocean drives the hydrologic cycle (Fig. 1-3).

The bulk of the remaining 3 percent of the earth's water, nearly all of which is fresh water, is inaccessible, with almost 2.25 percent of the 3 percent being locked up in snow and ice. A significant part of the remainder is contained in groundwater sources at depths of more than 2500 feet beneath the surface of the earth.[13] The end result is that we have relatively easy access to less than 0.8 percent of the earth's total water supply: 0.77 percent in the form of groundwater, 0.02 percent in the

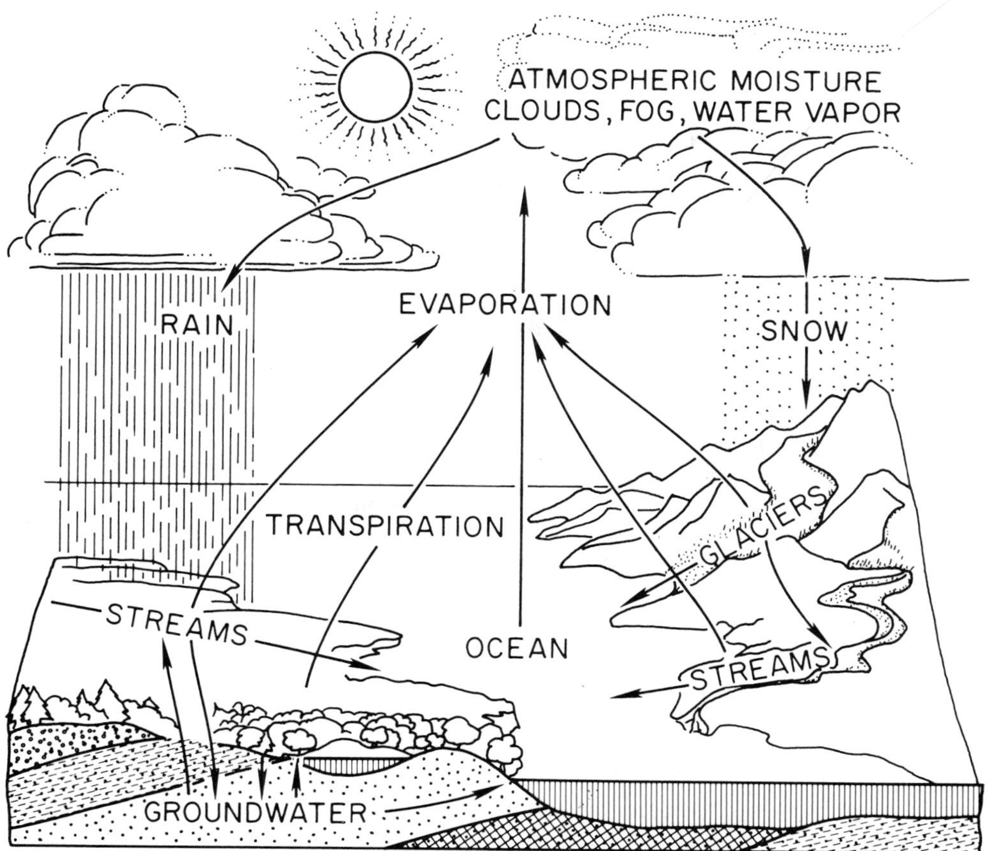


Figure 1-3 The hydrologic cycle.

form of surface water in lakes, and inland seas, and 0.0001 percent in rivers and streams. The atmosphere contains only 0.001 percent of the earth's total inventory of water.

It has been estimated that we could withdraw up to 5 million trillion gallons of water per year for global use without having saltwater intrusion into groundwater sources. Approximately 30 million trillion gallons of water return to earth each year in the form of snow or rain. Roughly 10 million trillion gallons of this ends up in streams and rivers. The remainder evaporates, seeps into groundwater sources, or is used by plants. Thus, the total freshwater available each year for use worldwide is of the order of 15 million trillion gallons.[14]

It is instructive to compare the amount of water available with that required to meet the needs of an industrial society such as the United States. During the 1930s the design value of 99 gallons of water per person per day was used in estimating our water needs. The current value is 151 gallons per person per day.[15] Of this amount, almost 40 gallons per capita are used each day directly in the home, as shown in Fig. 1 - 4.

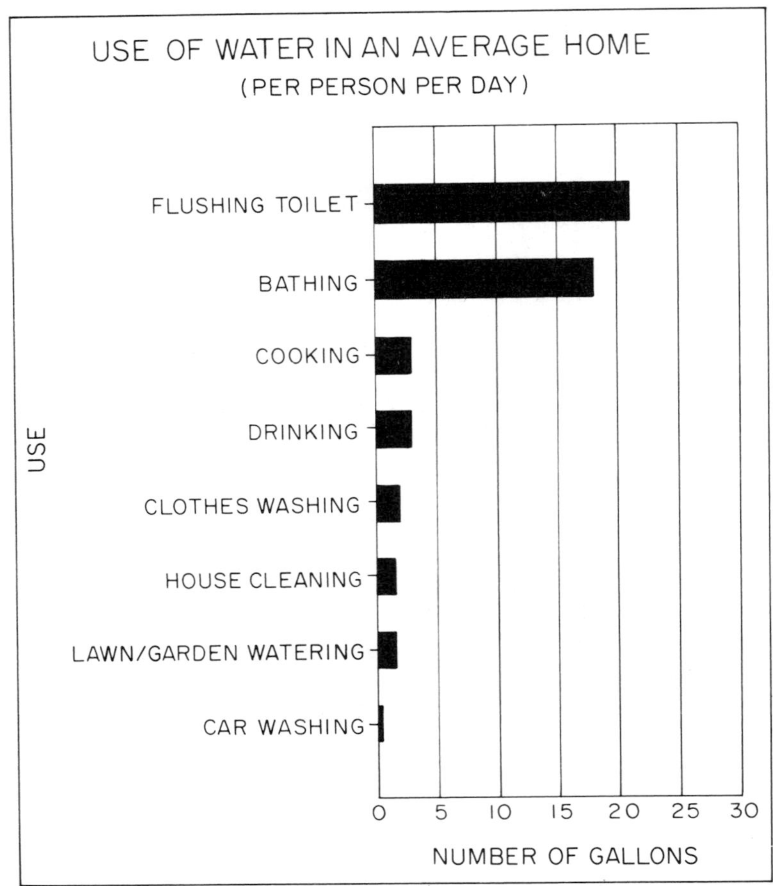


Figure 1-4 Use of water in an average home

---

Figure 1-5 illustrates the growth in the rate at which the United States is withdrawing water for all uses. It is projected that by the end of this century the total demand for water by the United States will exceed 1 trillion gallons each day and that this sum will be 50 percent more than can be provided from all our natural sources combined.[16] With this burgeoning demand for clean water, it becomes vital to understand the ways in which water supplies are being polluted by society's activities, including its solid waste disposal practices. This topic is explored in Chapter 4.

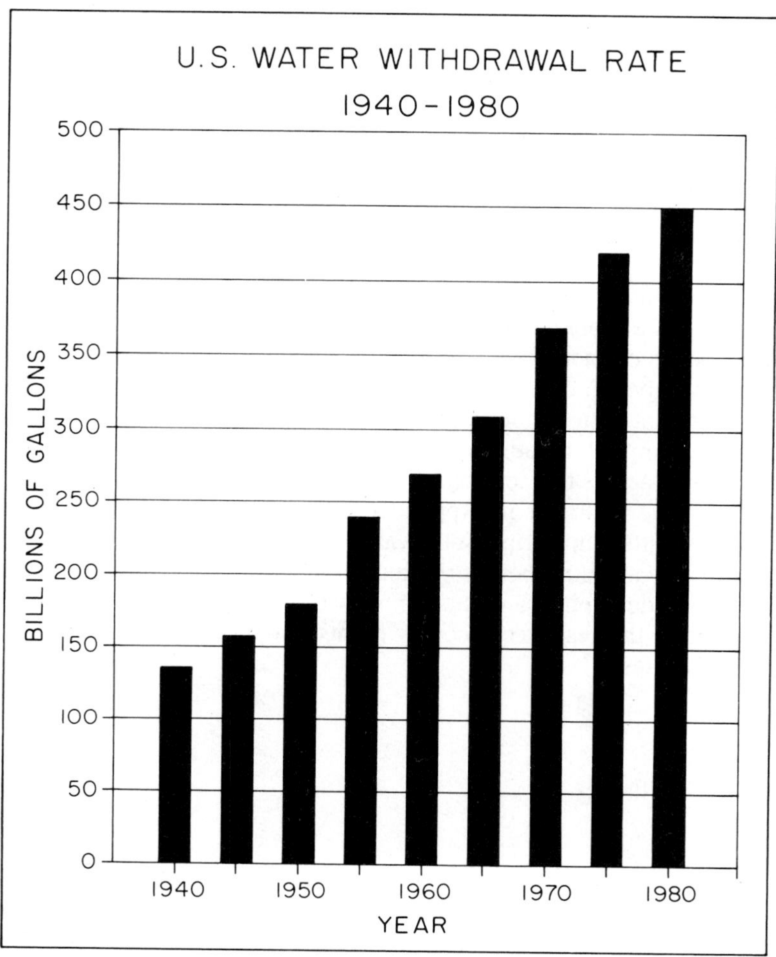


Figure 1-5 Rate of water withdrawal in United States.

## 1-5 SCOPE OF TEXT

In the ensuing chapters we describe the magnitude and complexity of the solid waste problem, review the various issues critical to understanding the management options, and assess the environmental, economic and public health impacts of each of the principal solid waste management alternatives.

Our purposes in writing this monograph are to make the reader aware that the solid waste management problem is an enormous and complicated one; that solid waste disposal will continue to be a persistent problem; that the solid waste management problem is growing in magnitude and complexity because of increased volumes of wastes and their changing character; that the management alternatives available to us are limited in number and that none is absolutely ideal. Each has advantages and disadvantages and their relative weights vary from one region of the country to the next even if the population, population density, and the volume and character of the wastes are the same. There is no universally "best" solution. The choice must be made on a case by case basis.

We have attempted to anticipate some of the readers' questions about solid waste and to answer them as objectively and completely as possible, given the existing state of knowledge. The answers we have provided are perhaps less important than the questions we have raised, questions we hope the reader will raise and demand answers to when his or her community seeks to determine how it will deal with its municipal solid waste problem. Because the generation of solid waste is and will continue to be a persistent activity, carefully planned and coordinated research programs should be developed and carried out to reduce the uncertainty associated with different options so that better choices can be made in the future.

Finally, we have attempted to summarize in Appendix 1 how a variety of municipalities and regions across the country are dealing with their municipal solid waste problems. We have included in Appendix 2 a summary of those federal regulations which directly affect municipal solid waste management, and Appendix 3 contains a summary of a symposium held in January, 1986 at the State University of New York at Stony Brook devoted to a discussion of waste management issues. Appendix 4 contains a list of the resource recovery plants in operation or under construction in the United States.

Fig. 2-1 Growing use of plastics by the American consumer.

# CHAPTER 2

## GENERATION OF GARBAGE AND TRASH

### 2-1 HISTORICAL ASSESSMENT

As nomads, humans had few waste disposal problems. If the food ran out or the garbage pile got too high, the individual or the tribe simply packed up and moved on. There seldom was any need to worry about refuse disposal; the garbage was left behind. And since early human garbage was "natural", it decomposed rapidly to melt into the natural background.

#### Debris Accumulation

Difficulties of debris accumulation began when humans established permanent residences, and became aggravated when the residences coalesed into towns. In ancient Troy every home had its own "garbage disposal unit" --the floor. Waste was simply thrown onto the floor or into the streets. The operative mechanism was gravity. When the floor had risen so high because of garbage deposition that headroom became limited, the doors and ceilings were raised to maintain living space and to extend the useful lifetime of the house and its "disposal unit".[1]

Regardless of whether garbage and trash are disposed of as in ancient Troy or in tidy dumps, in archaeological terms each generation of a city's inhabitants leaves behind a layer of waste. Over time, the accumulation of successive layers of waste results in an increase in the elevation of a city, and at a rate that may be significant. The solid waste accumulation rate of Bronze Age Troy was approximately 4 - 6 feet per century--more than 4.8 inches per decade![2] Surprisingly, modern day Manhattan Island would increase in elevation at a similar rate, if all its municipal solid waste were spread uniformly over the island.[3]

Normally people do not think of their wastes as contributing to the rise in elevation of their city. Indeed, the more common concept is one in which garbage is pictured as contributing to one isolated pile which, after a while, somehow decays and vanishes. Not so. The decomposition adds to the soil, but not everything decays. Further, although a single pile may receive the garbage and trash for a time, another pile (landfill) is soon required, then another, and another, leading to a gradual growth of multiple areas. Thus, there is growth in an increasing number of isolated areas. But the isolated areas resolutely refuse to remain isolated. Many forces act to disperse garbage from isolated piles. They include wind and erosion. Even the lazy citizen who can not be bothered with the proper disposal of his waste contributes to the spread of waste over large areas. Through such accretion, aggregation and dispersal of wastes the average elevation of the entire city is increased.

For areas inhabited for millennia, the effects of debris accumulation are easy to document. For example, the top of a tower erected about 7000 B.C. in the ancient city of Jericho is now almost below average "ground level" because of the build-up in the general area from debris accumulation. Indeed, evidence suggests that between 9000 and 1300 B.C. the average elevation of Jericho rose by nearly 70 feet![4]

Clearly, not all of the growth in elevation of a city is caused directly by garbage disposal. Natural disasters, such as floods and earthquakes, as well as rebuilding after significant events such as wars and fires also can affect the vertical growth of a city. But, in most of the sites which have been studied, the bulk of the accretion apparently results from the daily routine process of solid waste disposal. And vertical growth is not a comprehensive measure of garbage production. There also is a spreading effect in which urban areas grow laterally.

The fact that the vertical growth of cities has been more or less constant over the millenia with, for example, ancient Troy and modern day Manhattan having almost equal average rates of vertical rise, may indicate something about the rate of debris accumulation which people are willing to tolerate. A much more rapid rate of deposition presumably would not be acceptable aesthetically, or for other reasons. A much slower rate would indicate that extraordinary steps had been taken to dispose of debris or to limit urban development, and these actions evidently are not desirable economically or socially. There is much that we can learn about ourselves, our ancestors and other cultures from an analysis of garbage and trash, and of the ways in which we have disposed of it. Gunnerson points out that "debris accumulations in towns and cities are not all undesirable since they make life possible for archaeologists", and thus provide a special type of window on our past.[5]

## Garbage Disposal in Early Times

The very cities we look back upon as being extraordinarily enlightened, and as the forerunners of the customs and values of modern society, handled their garbage disposal in a very crude manner. It was not until 500 B.C. that the Greeks developed "municipal dumps", the first in the Western world. Ancient Rome had only open dumps to accommodate the refuse from its massive population. In 1400 A.D. the mounds of garbage beyond the city gates of Paris were so high they blocked the sighting of potential enemies, compromising the defense of the city.[6]

The Industrial Revolution marked a surge in the aggregation of humans into centers of great population density. The invention of devices such as the steam engine and weaving machines, along with the understanding of how to use coal to fuel machines, led to the creation of an enormous number of jobs at centralized facilities, making it attractive for many families to leave rural areas and move to the city. London was a classic example of such a city. The crowding of large numbers of people into relatively small quarters created environmental chaos. The standard practice for garbage disposal was to toss it into the streets or into the waterways. As a result, the streets reeked and the River Thames was made abominable.

Similar difficulties afflicted large United States cities as early as the 1600s and surprisingly crude conditions persisted well into the 19th century. As recently as 1842, Charles Dickens related experiences of his visit to New York City where he observed pigs roaming Broadway rummaging and rooting through garbage. The role of swine in municipal waste management was recognized as being helpful and, indeed, some cities had laws protecting pigs and other garbage-eating animals.[7]

In 1795 the Corporation of Georgetown (Washington, D.C.) enacted a law prohibiting the disposal of garbage in the streets. This law required individuals either to carry away their garbage themselves or to hire a private carter and was a significant advancement in changing the pattern of garbage disposal in our nation's capital. In 1800 President John Adams hired a private carter to haul garbage away from the White House, but it was not until 1856 that Washington had a city-wide garbage collection system supported by taxes.[8]

The degree of urbanization of a country provides some measure of challenges it faces in dealing with its solid waste disposal problems. For purposes of definition, an urbanized area consists of a central city or a central core, together with contiguous, closely settled territory, that combined have a total population of at least 50,000. Between 1840 and 1920 the urban population of the United States increased by nearly 30 fold: from 1,845,000 to 54,000,000.[9] The number of urban areas grew from 131 to 2,722 during this period, setting the stage for the United States to experience many of the urban environmental problems that had already beset European cities.

Cities did not take aggressive steps earlier to deal with their garbage disposal problems primarily because it was not until the 1800s that direct causative links were established between specific diseases and pathogenic organisms in wastes. The germ theory of diseases is a twentieth-century development, and it has provided a solid basis for justifying attempts to achieve environmental sanitation. Many people believed that there was a direct link between sanitation and health much earlier, but these views could not be substantiated unequivocally until the twentieth century.

Each seemingly significant advance by society has been accompanied by sobering side effects. Before the widespread adoption of the automobile, the principal means of transportation was the horse and carriage. In the early 1900s New York City had over 120,000 horses. Every day they produced 1,200 tons of manure and at least 60,000 gallons of urine.[10] The good old days! The horseless carriage alleviated the extensive health problems created by these animal wastes, but now we must contend with the problem of increased air pollution created by the combustion products of the fuel.

One of the first municipal incinerators in the United States was built in Allegheny, Pennsylvania in 1885. This new approach to reducing the amount of garbage was viewed positively as a sanitary way both of doing away with garbage and of simultaneously generating energy. Time revealed, however, that the designs of the early incinerators were flawed, that their operation was not cost effective, and that unless special steps were taken, the emissions from the incinerators posed a health problem.

Another popular way of disposing of garbage was to feed it raw to pigs. This apparently sound idea for recycling organic material was stopped in the United States in the mid-1900s when it was learned that the spread of vesicular exanthema disease (a disease of swine similar to the dreaded hoof and mouth disease) could be traced directly to the practice.

Among the most important lessons to be learned from an historical review of how society has dealt with its garbage and trash are that solid waste management is one of the most complicated, persistent and challenging problems ever faced by humankind, that it is unlikely there will be a simple all-encompassing solution best for all situations, and that we must persist in our search for new and innovative solutions.

## 2-2 CONSTITUENTS OF WASTE TODAY

A snapshot analysis of a sample of solid waste from a particular California community illustrates that the waste stream consists of a wide variety of items: discarded household and cleaning products (40 percent), automotive products (30 percent), personal products (16 percent), paint and related products (7 percent), and insecticides, pesticides, and herbicides (2.5 percent).[11] Indeed, essentially everything we find in the home or workplace ends up, in some quantity, as a component of our solid waste. The items are organic and inorganic; natural and anthropogenic; innocuous and toxic and hazardous. Among the items we find in the home classified as toxic or hazardous wastes are butane fuel, shaving lotion, insect killers, paints, disinfectants, hair sprays, copier inks, motor oils, fabric softeners, detergents, bleach, shampoo, drain openers, dishwashing soaps, window cleaners, hydrogen peroxides, liquid waxes, metal cleaners, glazing compounds, and stain removers. In the aggregate, the amount of these materials is staggering!

The magnitude of future solid waste disposal problems will be determined by a number of factors, including the amounts of non-consumable, disposable goods people acquire, the chemical makeup of these goods, and the willingness of society to recycle products. These in turn depend on the affluence of society, on the population, on packaging practices, and on scientific and technical advances. The success of recyling will depend upon the courage and conviction of our political leaders in assessing the full costs of all disposal options; in eliminating, or at least reducing, through legislation the economic incentives for using virgin materials; and in enacting and enforcing other important, but unpopular, legislation with appropriate economic incentives and disincentives.

According to a survey conducted by Franklin Associates, Ltd. the nationally averaged municipal solid waste composition in 1980 is as shown in Fig. 2-2.[12] Over 79 percent by weight of municipal solid waste is combustible. The most abundant components of the combustible fraction are paper (29.8 percent of the total), yard waste (19.1 percent) and food (16.7 percent). The most abundant component of the non-combustible fraction is glass (10.5 percent of the total) and metals (9.3 percent). Further analysis reveals that more than half of the solid waste is from paper and packaging. More than 50 percent of the aluminum, 25 percent of the iron, and 50 percent of the glass in the waste stream comes from soft drink and beer containers.[13]

### Paper Products and Cellulose

A major constitutent in trash is cellulose, the principal substance in the cell walls of trees and most plants and vegetables. Cellulose, a carbohydrate, is composed of a specific combination of carbon, hydrogen, and oxygen. Wood products contain over 40 percent cellulose, cotton over 95 percent and paper is almost 100 percent cellulose. Because cellulose is highly combustible and constitutes such a large fraction of municipal solid waste, incineration of solid waste is very practical. The heats of combustion of paper products and of a number of other energy sources expressed in

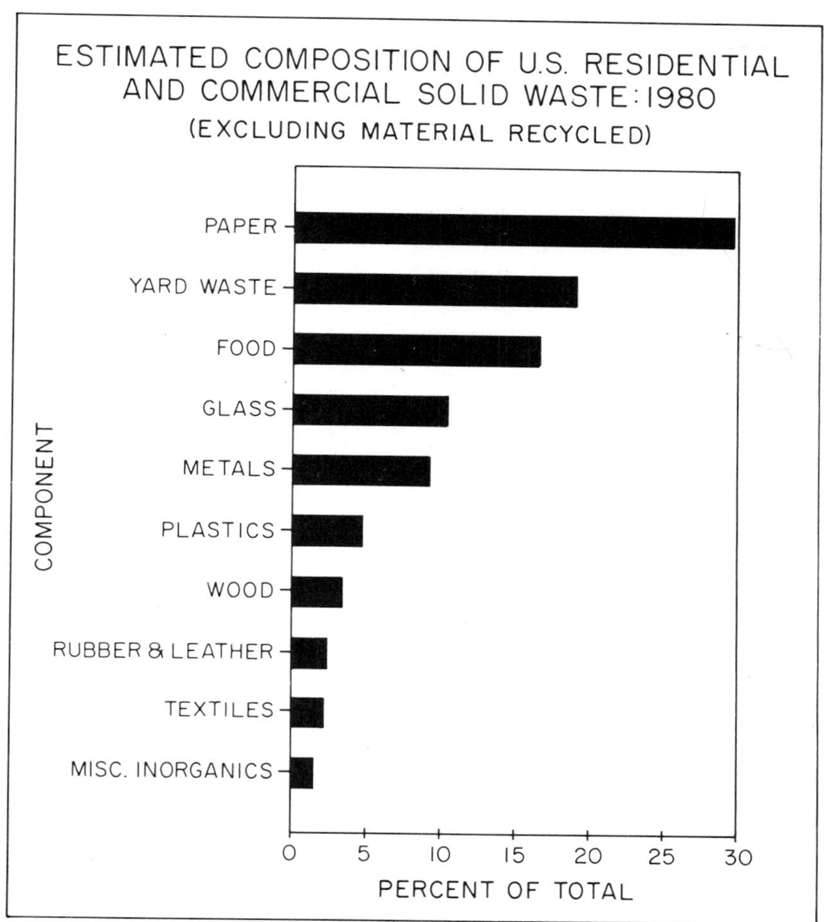


Fig. 2-2 Estimated composition of solid wastes generated in United States.

---

BTUs per pound of oven-dried material are as follows: newspaper 5,980, magazines/books 5,810, corrugated cardboard 5,600, and wood 8,600. For comparison, the corresponding value for coal is 6,300 to 14,500, charcoal 12,000 to 13,000, fuel oil (low sulfur) 19,000 to 19,300, and gasoline 21,000.[14]

Paper products are all rather simply related to chemical structures found in nature. This is unlike the case of numerous products which have resulted from the advances in synthetic organic chemistry described in the next section.

## Synthetic Materials

Organic chemistry is the study of carbon compounds, particularly those which link carbon atoms together in combination with hydrogen, oxygen and nitrogen. Such compounds are formed naturally by living systems--a fact which stimulated the first studies. Our fossil fuels, coal, oil and natural gas, were formed from natural organic compounds over long periods of time as a result of geological processes. Today we have learned to synthesize not only many of the natural compounds but countless others which are more or less "foreign" to the environment.

While the hydrocarbon chains which occur in coal, gas and petroleum are complex, because they have been present in the environment for millions of years they have been encountered and are "recognized" by plants and animals and under most circumstances do not present significant adaptive environmental problems. Through modern chemical techniques, however, we have produced synthetic organic compounds of much greater complexity which do not occur naturally, and which plants and animals have had no experience in processing. According to one source, the total United States' production of synthetic organic chemicals increased from 1 billion pounds in 1940 to 300 billion pounds in 1976.[15] This phenomenal growth is associated with products such as nylon, rayon, polyethylene, polystyrene and with the numerous plastics that are used extensively in all aspects of our daily lives, from the packaging of food to the fabrication of electrical and mechanical devices, including artificial hearts.

Even more unusual and persistent materials can be formed by adding chemical elements called halogens to the hydrocarbon chains. With the additions to hydrocarbons of elements such as chlorine, bromine, or iodine (the halogens), materials such as perchloroethylene (PCE) and trichloroethylene (TCE) have been developed as industrial solvents. Similar compounds have been developed as pesticides and drugs.

By adding chlorine to specific hydrocarbons it is possible to fabricate the polychlorinated biphenyls, commonly referred to as PCBs. Because these chemicals are particularly good electrical insulators, they found extensive applications in high voltage electrical machinery such as transformers. Other variants of the halogenation process result in the production of wood preservatives and the polyvinyl chlorides (PVCs) used in a variety of plastics, including phonograph records and plastic pipe.

The halogenated hydrocarbons are persistent in nature and can alter natural biochemical and biological processes. Their efficacy as pesticides results from their persistence and their ability to interfere with essential life processes. They are relatively insoluble in water, but highly soluble in fats, and thus tend to accumulate in the fatty tissue of higher organisms at concentrations greater than in the surrounding environment. This point is of particular interest because many members of the halogenated hydrocarbon group are known or suspected carcinogens.

The persistence of halogenated hydrocarbons is an issue of concern for landfilling and for incineration. Disposal of halogenated hydrocarbon products in landfills can lead to the release of dangerous material as leachate. Natural microbiological processes alter such materials, but on time scales much longer than with less complex chemicals. In the decomposition process, the hydrocarbons may be transformed into less dangerous forms, but there is also concern that other even more toxic forms may be produced.[16] Although halogenated hydrocarbons are relatives of the petrochemicals which we utilize for fuels because of their combustion properties, the addition of the halogens to municipal solid wastes makes combustion much more difficult, since temperatures in excess of 900° Fahrenheit are required to break the chemical bonds of many of these compounds. There is concern that at lower temperatures large amounts of halogenated hydrocarbons could be released into the atmosphere and affect the health of individuals in the vicinity of the incineration plants. This is one of the reasons why the combustion temperature in incinerators is important.

## 2-3 QUANTITY OF WASTE PRODUCED; FUTURE PROJECTIONS

The complexity of garbage disposal issues is closely coupled to population growth and urbanization. If we lived in a world where each family had hundreds of acres for its use, there would be no difficulty in reserving a small portion of each estate for the disposal of wastes. But we do not. Instead of hundreds of acres per family, we must contend with situations in which we have hundreds of families per acre. Moreover, the economics of solid waste disposal are such that only relatively small outlays can be made to insure that the garbage generated is disposed of properly. Options such as trucking garbage thousands of miles to a remote part of the country are not feasible. Because of the influence of urbanization on solid waste management policies and practices, this section is devoted to a review of population trends.

From a global perspective, over one-fourth of all the people ever born are now living. There has been an enormous growth of the world's population in the last 2000 years (Fig. 2-3). The population explosion has occurred partly because of major advances in the control of diseases and partly because of expanded food sources: both have resulted from technical advances. A mushrooming population and increasing demands and expectations concerning the quality of life have brought tremendous pressures on traditional methods and mechanisms for disposing of garbage and trash.

In the year A.D. 1 the world population was roughly 250 million. By A.D. 1850 it had reached roughly 1 billion. It is now over 4 billion, and is projected to grow to between 6.5 and 7 billion by A.D. 2000.[17] Of the 4 billion people who now inhabit the earth, over 40 percent are under the age of 15. The stage is thus set for a continued mushrooming of the world's population. The demands of this growing population for food, transportation, shelter, and other items will add to the problems of solid waste collection and disposal.

In the United States the population in 1850 was slightly more than 23 million. By 1900 it had grown to 96 million; by 1950 to 152 million.[18] Now it is 238 million, and the projection for the year 2000 is 268 million[19] (Fig. 2-4). Although our country is not expected to realize a rate of population growth as high as many other regions of the world, the steady increase that will occur will be significant and will impose major challenges to those who must plan for the delivery of public services, including municipal solid waste collection and disposal.

Because of the special problems associated with solid waste management in densely populated regions, it is instructive to examine the rate at which the United States is becoming urbanized. Table 2-1 shows that more and more of our citizens are leaving rural settings and moving to cities. This trend, as well as the overall population growth, must be taken into account explicitly in all planning efforts.

One driving force, and one indicator of garbage generation, becomes evident from an examination of the energy use in the country. Figure 2-5 shows that near the turn of the century energy usage was at the level of 300 million billion kilocalories per year.[20] By 1975 it had risen to over 1,400 in these same units. Inasmuch as energy is required in virtually every aspect of the running of industry, from the

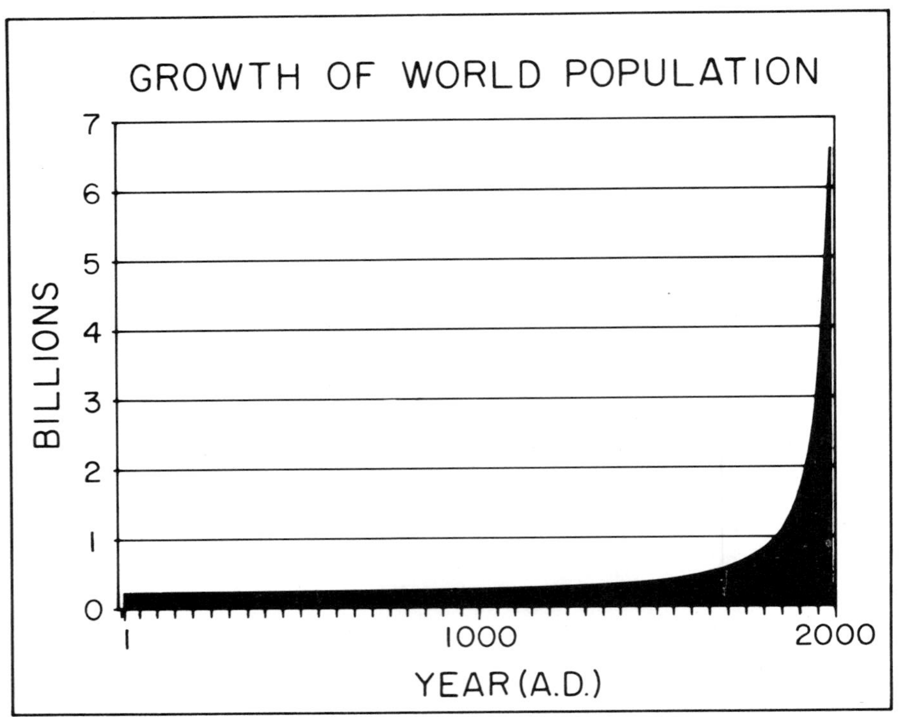


Fig. 2-3 Growth of world population.

extraction of raw materials to the manufacturing of chemicals and new plastics, it represents a measure of the production of products which will eventually become trash.

In view of such considerations, it is estimated that the solid waste generated in the United States requiring disposal will grow by an average of 1.8% per year between 1980 and 2000. Approximately two-thirds of this amount will be due to population growth, while one-third will be due to the increased amount of garbage and trash generated on the average by each person.[21] One must keep in mind that the population figures mentioned above are projections for the nation as a whole, and that individual regions can grow at much greater, at much smaller, or even at negative rates. The present distribution of population in the United States is shown in Fig. 2-6. The striking concentrations of populations in the Northeast and the Pacific Southwest are clearly evident in this three dimensional representation, which graphically illustrates the immense solid waste disposal problem which will be faced by various regions of the country.

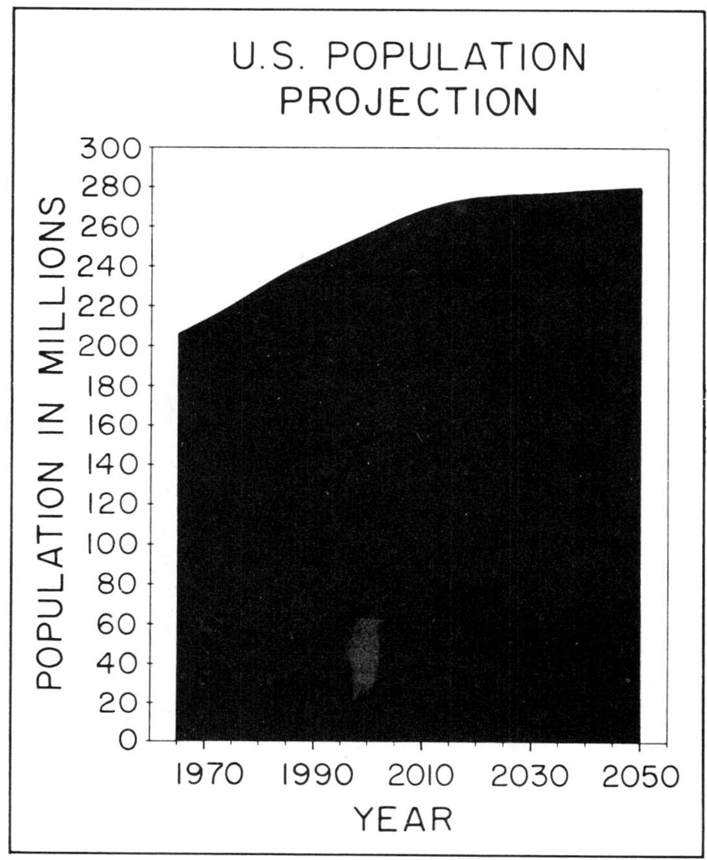


Fig. 2-4 Projection of future population growth in United States.

---

Figures 2-7a and 2-7b illustrate the projected total annual solid waste generation rate for the United States, and the per capita annual rate of solid waste generation. By the year 2000 it is projected that we will be disposing of over 180 million tons of garbage and trash per year, or approximately 3.7 pounds per day per person -- more than 1300 pounds per person per year. Note that both the total and the mass per person are expected to grow in the United States for the remainder of the century.

## 2-4 RECYCLING

Clearly, one way to reduce the amount of garbage and trash to be disposed of is to reclaim products that can be reused in the manufacture of new consumer products. This concept, called "recycling", is definitely an important one, but it has certain limitations.

**Table 2-1 Distribution of United States population by size of place.**

| POPULATION OF UNIT | Percent of total population by year | | | | | |
|---|---|---|---|---|---|---|
| | 1800 | 1900 | 1920 | 1940 | 1960 | 1980 |
| Greater than 1 million | - | 8.4 | 9.6 | 12.0 | 9.8 | 7.7 |
| Between 500,000 and 1 million | - | 2.2 | 5.9 | 4.9 | 6.2 | 4.8 |
| Between 250,000 and 500,000 | - | 3.8 | 4.3 | 5.9 | 6.0 | 5.4 |
| Between 100,000 and 250,000 | - | 4.3 | 6.1 | 6.0 | 6.5 | 7.5 |
| Between 50,000 and 100,000 | 1.1 | 3.6 | 5.0 | 5.6 | 7.7 | 8.7 |
| Between 25,000 and 50,000 | 1.3 | 3.7 | 4.8 | 5.6 | 8.3 | 10.3 |
| Between 10,000 and 25,000 | 1.0 | 5.7 | 6.6 | 7.6 | 9.8 | 12.2 |
| Between 5,000 and 10,000 | 1.8 | 4.2 | 4.7 | 5.1 | 5.5 | 6.8 |
| Between 2,500 and 5,000 | .9 | 3.8 | 4.1 | 3.8 | 4.2 | 4.1 |
| TOTAL URBAN* | 6.1 | 39.6 | 51.2 | 56.5 | 69.9 | 73.7 |
| TOTAL RURAL | 93.9 | 60.4 | 48.8 | 43.5 | 30.1 | 26.3 |

Source: U.S. Bureau of the Census, *Statistical Abstract of the United States:*1985 (105th edition.) Washington, DC, 1984

*An urbanized area consists of a central city or a central core, together with contiguous, closely settled territory, that combined have a total population of at least 50,000 (United States Bureau of the Census, 1984. Statistical Abstract of the United States: 1985 (105th Edition). Washington, D.C.

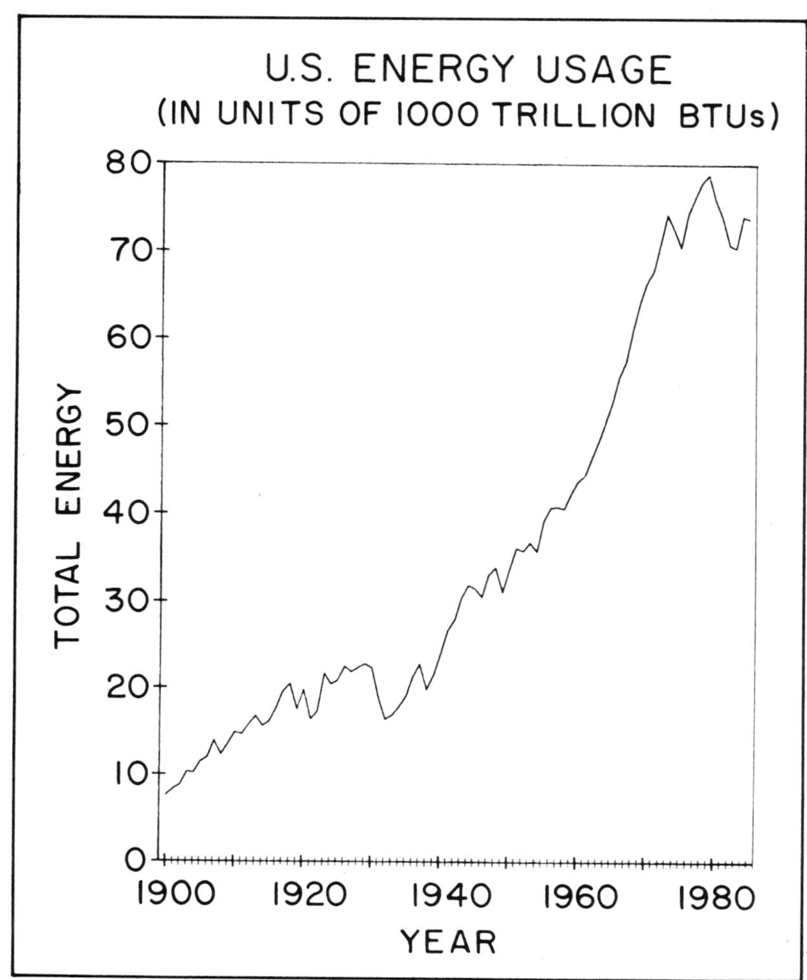


Figure 2-5 Energy usage in the United States.

---

The potential for recycling stems partly from the fact that an estimated two-thirds of the materials we now discard could be reused. According to the Worldwatch Institute, "With products designed for durability and for ease of recycling, the waste streams of the industrialized world could be reduced to small trickles."[22] But, recycling efforts in the United States have been only marginally successful.

Approximately 50 percent of municipal waste can be classified as paper products, and the reclamation process for most paper products is relatively straightforward. Public concern for environmental matters represents a strong force in favor of recycling, but the low cost of paper products produces little economic incentive for recycling. The situation is somewhat better in the recovery of metal products.

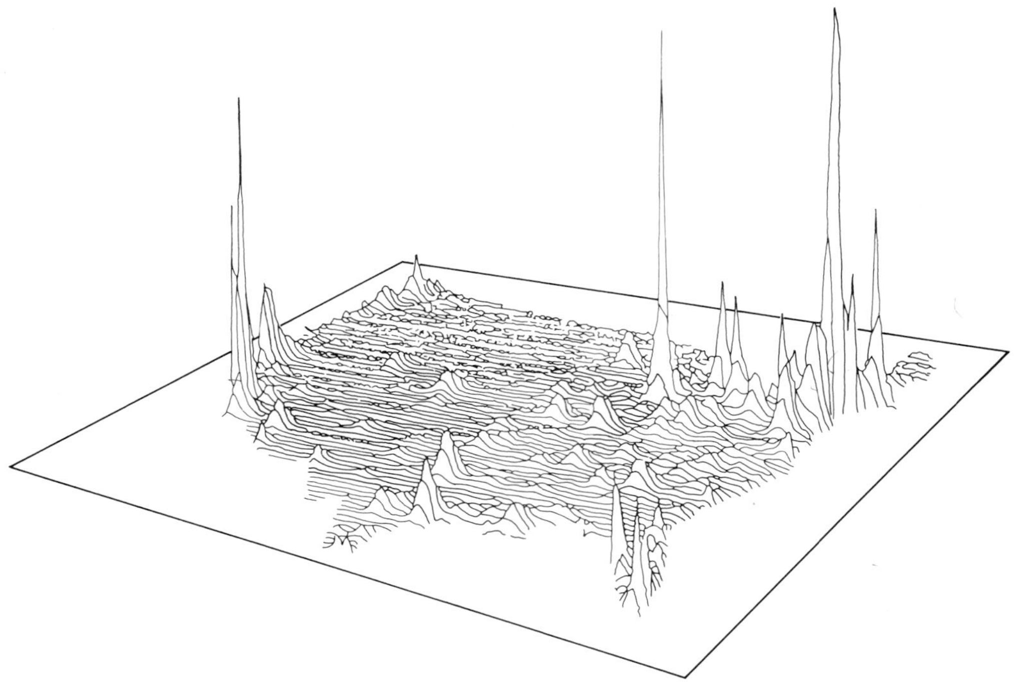

Figure 2-6 Distribution of population in the United States by region

The present use of paper in the United States currently is 600 lbs per person per year. Less than 20 percent of it is recycled, in contrast to a recycling rate of about 35 percent during World War II. Economic considerations strongly influence current practice. It is cheaper for the paper manufacturer to harvest virgin trees, transport them to a paper factory and manufacture the final paper products than it is to haul wastepaper to a plant for recycling and processing. Morever, there is a consumer preference for "new" products, which reduces even further the incentive for a company to engage vigorously in recycling operations.

There is a degradation in the quality of recycled paper and thus recycling is of use only in particular applications. Recycling of paper causes the fibers to be shorter and weaker. Moreover, the innermost packaging of foodstuffs--a significant use of paper products--must use new paper material because federal laws prohibit the use of material which might have come into contact with toxic substances.[23]

There are many other factors, primarily socio-economic in nature, that influence the paper recycling rate.[24] Most paper mills are located in rural areas, near the trees. For a major change in the economics of paper recycling to occur, paper mills would need to be near population centers where most of the waste paper products are generated. Shifting paper mills away from rural areas would have a devastating impact on the local economies, since in many cases the mills are the primary employers. Thus, natural forces will not automatically bring about a substantial shift toward recycling. A far-sighted national policy, taking into account the overall economic impact, would be required.

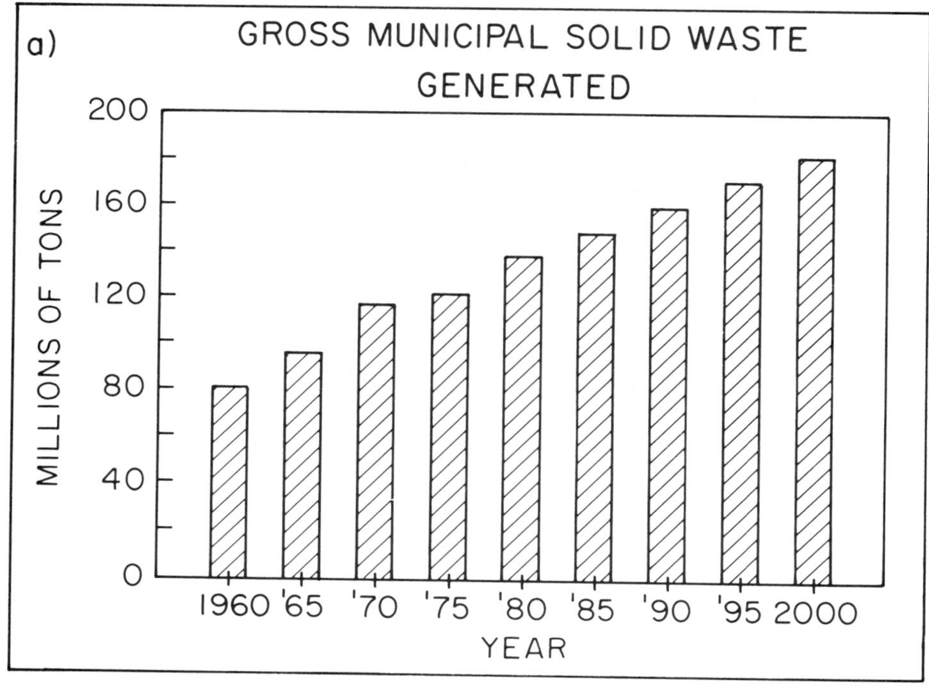


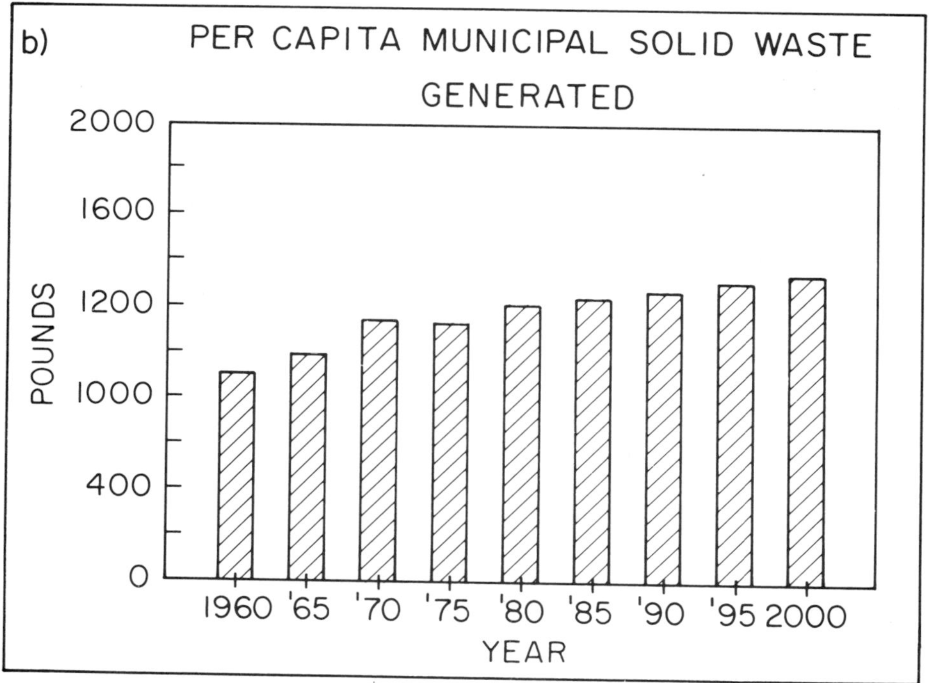


Figure 2-7 (a) Projected annual solid waste generation rate for the United States; (b) per capita annual rate of solid waste generation.

Incineration in efficient resource recovery facilities is another possible option for recovering the energy content of paper. When equal weights of coal, oil, and paper are burned, paper produces about half the energy of coal and a third the energy of oil.[25]

The recycling rate of metal is more than double that of paper. Ferrous (iron) materials can be magnetically separated from waste streams, melted down, and reused. Aluminum cans, the primary containers for beverages in the country, are being reclaimed at a significant rate, aided by deposit/return laws in many states.

Of the 1.1 billion pounds of waste (of all kinds) generated by Americans daily, roughly 230 million pounds are steel, 12 million pounds are aluminum, and 7.4 million pounds are copper. The aluminum recovered is about 60 percent of that produced, while the iron recovered is 12 percent of that produced. Of the 55 billion aluminum cans used in the United States each year, approximately 80 percent are recovered through consumer returns in states that have mandatory deposits. In addition, procedures have been developed for recovering aluminum from municipal refuse, by separating out ferrous and non-ferrous material from the ash of resource recovery facilities and using dense-media separation techniques for the non-ferrous portion. This is an effective process and has promise for substantially reducing the packaging costs for beverages, since the energy needed to recycle aluminum containers is much less than that required to produce the same amount of aluminum from bauxite ore. And there is no loss in quality. Unlike the case of paper where recycling leads to a loss of product strength, recycling of aluminum following incineration often produces an alloy with even greater strength than the original aluminum product because of the admixture of small amounts of other metals.

Other principal components of municipal refuse which lend themselves to recycling include plastics and glass. Neither of these materials, however, offers the possibility of producing a major reduction in the solid waste stream in the near future. Plastics pose a particularly difficult problem. They must be removed from the waste stream before incineration and segregated by chemical type into thermoplastics such as polyethyelene, polystyrene and nylon; and thermosets, such as polyesters, phenolics and epoxies. Glass, which is basically quartz sand, generally is not regarded as an environmental burden, and is relatively inexpensive to manufacture. This limits the pressure for any massive undertaking to recover glass products from wastes.

The subject of recycling is discussed further in Chapter 8.

Figure 3-1 Examples of vehicles used in collecting garbage and trash.

# CHAPTER 3

## COLLECTION AND TRANSPORTATION SYSTEMS

### 3-1 COLLECTION VEHICLES

Every working day, tens of thousands of heavy vehicles go from residence to residence and from company to company picking up residential and commercial solid waste for delivery to a landfill, a transfer station, or a resource recovery facility. Many people believe that once their garbage and trash is removed from their curb, dumpster or backyard, the problem is over. Last week's refuse is totally out of sight and, for many, out of mind once the garbage truck rumbles down the street. We know that, from an environmental standpoint, the problem is really just beginning. The emissions from the massive fleet of garbage trucks, the traffic congestion aggravated by the trucks, the littering caused by trash blown off the trucks, and the disposal processes themselves all contribute to environmental disturbances which would not have occurred if there were no wastes. But there are wastes and there always will be wastes. In this section we focus on some of the principal issues associated with collecting the garbage and trash. In subsequent chapters we deal with its disposal.

A typical garbage truck can hold 10 tons of garbage and can compress typical residential garbage into 30 percent of its curbside volume. It travels 20 miles a day, is diesel powered, and has a fuel mileage of 2 to 5 gallons per mile.[1]

Expenditures for garbage collection equipment are a major component of any municipal budget. New York City has 2,331 garbage trucks, all diesel powered. The average truck travels 15 miles per day, and must travel approximately 7 miles from the center of its pickup area to its unloading point at a landfill or transfer station. The average cost of a new truck is $60,000. Its annual maintenance cost is $12,000, and it requires approximately 1500 gallons of fuel per year. In 1984 the cost of collecting New York City's garbage and trash was $92 million, or roughly 80 percent of the total cost of collection plus disposal.[2].

Cities such as New York City may have to employ complicated methods to transport their garbage and trash from residences to the final disposal site. Trucks carry garbage from dumpsters to piers, where barges are loaded. The barges then travel to a landfill, where they are unloaded onto vehicles which carry the wastes to designated areas on the landfill site (Fig. 3-2).

### 3-2 ENVIRONMENTAL IMPACT OF REFUSE COLLECTION

It is difficult to quantify accurately the contributions of contaminants to the environment from refuse collection and disposal systems. Upper limits of vehicle emisions associated with collection can be estimated, however, since vehicle emissions are regulated by Federal legislation. Using the Federal standards, the total number of trucks on the road and the travel time per day, the maximum total emissions of regulated pollutants can be calculated. Actual emissions may be much lower.

Table 3-1 compares the projected emissions of selected pollutants for the resource recovery facility planned for the Brooklyn Navy Yard in New York City(described in Ref. 3) with emissions from the entire fleet of New York City's garbage trucks. Total truck emissions were calculated using (1) current EPA standards in grams/horse-power/hour with (2) an estimate of 200 horse-power for the average collection vehicle, (3) 15 miles of travel per day for each truck, and (4) 2,331 trucks moving at an average speed of about 4.5 miles per hour.[4] Particulate emissions were calculated from the EPA standards that will go into effect in 1988[5] and probably represent minimum values for particulate emission rates in 1985. The particulate emissions provided the basis for calculating the total tetrachloro-dibenzo-p-dioxin (TCDD) and 2,3,7,8,-TCDD emissions.[6] These compounds are included here because they are considered to be among the most toxic and carcinogenic substances known (see Chapter 6). Other contaminants also could be important. Because of the assumptions in the calculations, all total emission rates probably are correct to within about an order of magnitude (a factor of 10). This level of accuracy is sufficient for the purpose of our comparisons.

**Table 3-1. Projected pollutant emissions from the proposed Brooklyn Navy Yard resource recovery facility compared to emissions calculated for the entire fleet of New York City Department of Sanitation garbage trucks.**

| | Tons per Year* | | | | | |
|---|---|---|---|---|---|---|
| Source | CO | NOx | HC | TP | Total TCDD | 2,3,7,8-TCDD |
| Brooklyn Navy Yard | 366 | 2973 | 65 | 161 | $3.7 \times 10^{-5}$ | $2.4 \times 10^{-6}$ |
| Dept. of Sanitation Trucks | 6916 | 4774 | 580 | 268 | $6.2 \times 10^{-9}$ | $8.0 \times 10^{-10}$ |

* CO = carbon monoxide, NOx = total nitrogen oxides, HC = total hydrocarbons, TP = total particulates, TCDD = tetrachloridibenzo-p-dioxin.

Figure 3-2 (a) Barges being loaded with trash and garbage in New York City for transport to Fresh Kills landfill on Staten Island; (b) barges being unloaded at Fresh Kills landfill.

Figure 3-2 continued

Table 3-1 shows that total emissions of most pollutants from New York City's fleet of garbage trucks are greater than from a single large resource recovery facility such as that proposed for the Brooklyn Navy Yard--a facility which would handle about 11-12 percent of the City's trash. Differences between the two sources are not large. Projected dioxin emissions from a single resource recovery facility, however, are several orders of magnitude greater than those calculated for all of New York City's refuse collection vehicles.

Table 3-1 indicates that concern about emissions of dioxins in New York City should be directed primarily at the disposal end of the waste collection and disposal system.[7] From these observations, it appears that the best policy for controlling air emissions is to optimize economic aspects of refuse collection to help ensure the financial support needed to upgrade pollution control in incinerators and landfills where the more serious environmental problems may occur.

Most of the current trends in refuse collection practices in the United States are dictated primarily by economic incentives. Communities are moving toward more cost-effective refuse collection systems. High-volume, automated sideloading trucks and stationary containers are either being used or considered for use in many areas because they reduce labor, fuel, and maintenance costs.[8] The number of transfer stations, whose functions are described below, is increasing as distances to disposal sites get larger. Because most of these practices minimize truck travel time and hence emissions, strict Federal controls beyond emissions regulations probably will not be required.

It is unlikely that a community's choice of any particular waste disposal method will significantly affect total emission rates from refuse collection vehicles. The one exception might be if resource recovery facilities were constructed close to the source areas. Vehicle emissions are dictated primarily by total truck travel time, which is a function of routing, container location (curbside or back-of-the-yard), frequency of collection, and the use of transfer stations. The last of these factors has been shown to be particularly important for reducing truck travel in large cities.[9].

## 3-3 TRANSFER STATIONS

The optimal routing and scheduling of garbage trucks for the pick up and disposal is a complicated mathematical problem. Indeed, this generic problem has attracted the attention of many scientists and engineers in recent decades. An example of a question that might be posed is how should a city route its collection vehicles over its 2000 miles of streets to minimize travel distance and collection time to its two landfills and one resource recovery plant. A detailed solution of this kind of problem can be achieved for a set of prescribed conditions. An interesting finding from a general study of the problem is that the addition of an intermediate collection site--a transfer station--improves transportation efficiency. The principle behind the transfer stations is that a fleet of smaller trucks delivers garbage to the station, for transfer to a smaller fleet of larger vehicles which then transport it to the final disposal site. Normally there is some mechanism at the transfer station for additional compacting of the refuse.

Transfer stations are sites for the temporary storage of garbage and trash. They are located centrally so pick-up vehicles can dump their loads without having to make long trips to the countryside to unload at landfills or resource recovery facilities. Instead of spending time on long, unproductive runs, the standard vehicles can spend more time picking up more municipal garbage and trash.

The transfer station concept did not emerge from a recent sophisticated mathematical model. It was employed even during the days when garbage was collected by horse and cart. With the introduction of motorized vehicles, the relative advantage of intermediate collection sites diminished because the time for the trip from pick-up to dump was so short compared with previous practices. Once again, however, the pressures placed on the collection system are so great that the transfer station is back in vogue.

Figure 4-1 Fresh Kills landfill in New York City

# CHAPTER 4

## ENVIRONMENTAL IMPACTS OF LANDFILLING OF SOLID WASTE

### 4-1 INTRODUCTION

Open dumping is the age-old practice of discarding refuse at designated locations. It is the oldest method employed by humans to separate garbage and trash from their immediate surroundings. Starting in the Middle Ages, dumps often were set afire both to reduce the volume of the accumulated trash and to achieve what was thought to be a degree of purification associated with the burning process. With the exception of this innovation, landfilling has continued to be used essentially unchanged over the millenia.

In the past half-century the combined environmental impacts of smoke pollution, odor generation, attraction of pests, groundwater contamination and the other disadvantages associated with open dumps have led to the development of the technique known as sanitary landfilling. At a sanitary landfill, the refuse received each day is spread out in a thin deposit and covered that same day with a thin layer of soil. When some prescribed maximum height is reached, the landfill is covered with a few feet of soil, compacted and closed.

In many areas of the country this approach to garbage disposal is clearly appropriate. If population density is low, if land is available, if there are no vital groundwater or surface water sources near the landfill site and if the waste components are primarily biodegradable, it is difficult to imagine a more efficient and safer approach to municipal solid waste disposal.

Often, however, this is not the case. In many metropolitan areas landfills are nearing the ends of their useful lives; new sites are difficult to identify and nearly impossible to acquire. In other areas, runoff and leachate from landfills are contaminating surface and groundwater supplies. Because of these problems, alternatives to landfilling are being pursued vigorously in many parts of the United States.

The three methods of sanitary landfilling are trench, area, and ramp. Each method has advantages; the best method for a particular site depends on the topography, geology, and hydrology of the area. In all three methods refuse is dumped within specific areas, spread out, and compacted. The resulting refuse cells are covered with soil at the end of each day. When the landfill has reached the intended elevation, three to four feet of soil is usually applied as final cover.

In the trench method refuse is dumped into long narrow trenches. The excavated soil is used as cover material. In the area method solid waste is spread over the ground and cover material is usually brought from another location. The ramp method is a combination of the other two. A small excavation is made and refuse is deposited on the face of the resulting slope. The three methods are illustrated in Figs. 4-2 through 4-4.

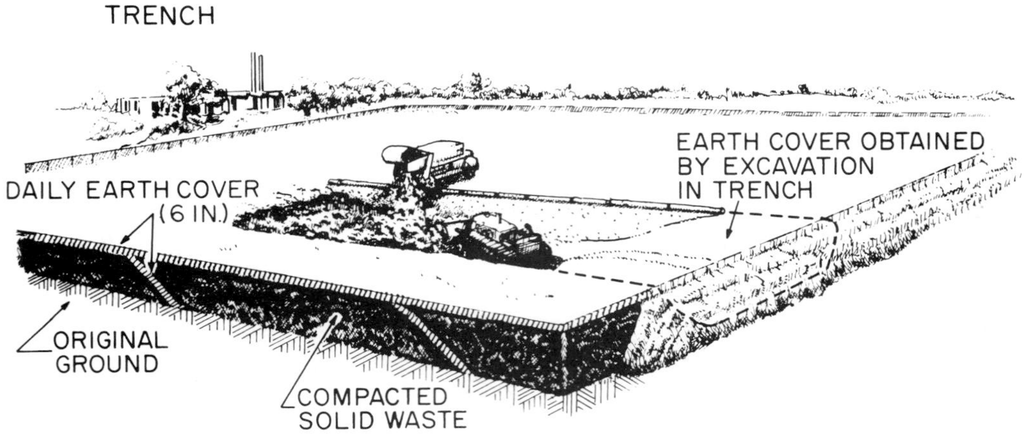


Figure 4-2 Trench landfilling method

The design and construction of a sanitary landfill is a major engineering project which should be undertaken with the same care and attention given any large technical project. The proper investment of effort in the siting of the landfill and in analyzing potential environmental impacts is required from the outset. Basic information such as the waste generated per capita over approximately a 30-year period, the chemical makeup of the waste, and the hydrogeologic nature of a variety of sites should be carefully weighed in selecting a site. Special efforts should be made to involve from the outset broad community participation in discussions of all aspects of siting, design and operation of a landfill. Sociological factors and perceptions are of major significance.

In the following sections we examine some of the considerations that should go into the design and construction of sanitary landfills and discuss some of the advantages and disadvantages of landfills.

## 4-2 SURFACE WATER

Surface water provides over half of the water needs of the entire United States.[1] Roughly 20 percent of the annual rainfall received on land ends up in storage in, or en route to, lakes, reservoirs, and the oceans. Considerable attention must be given to treating water from lakes and streams before human use. Since numerous municipalities typically are located along the banks of most rivers, the discharge from one community often becomes the source for another further downstream.

Figure 4-3 Area landfilling method

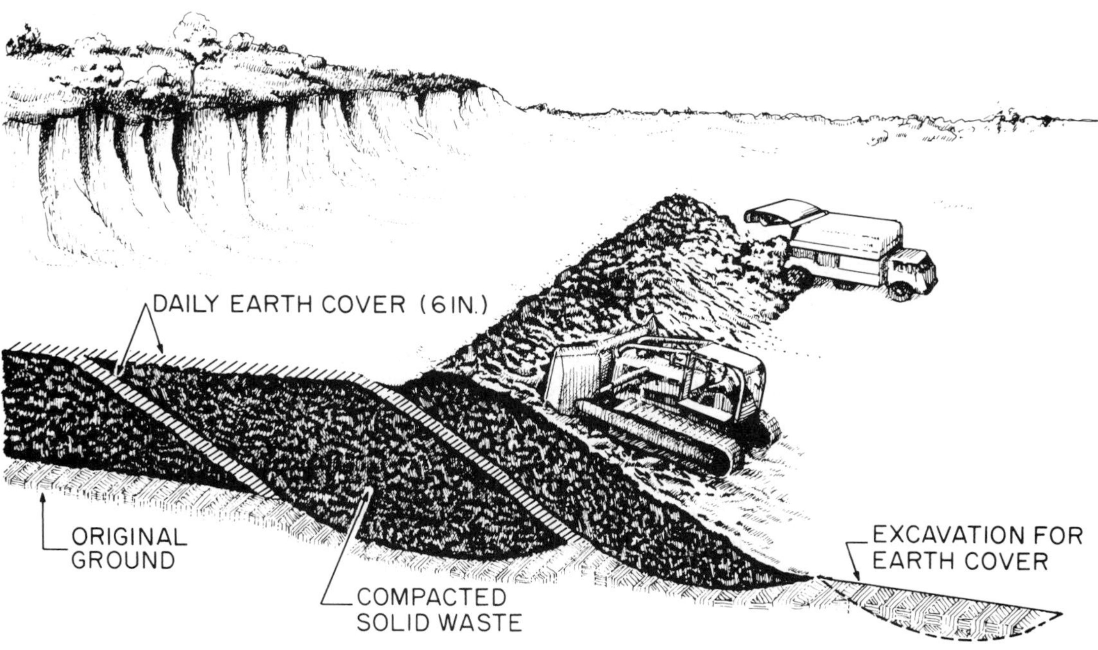


Figure 4-4 Ramp landfilling method

In some areas of the country where groundwater is not available, provisions have to be made to bring water to metropolitan areas from great distances. Examples include Los Angeles and New York City. In the case of New York, water must be drawn a distance of about 100 miles from reservoirs in the Catskill Mountains and from the Delaware watershed (Fig. 4-5).

The situation is even more severe in the greater metropolitan area of Los Angeles (Fig. 4-6). The City of Los Angeles receives by aqueducts about 80 percent of its water from snow melt in the High Sierra mountain range about 340 miles away, 15 percent from local groundwater, and 5 percent from the Metropolitan District which serves the suburban area outside of Los Angeles. The Metropolitan District takes its water from the Feather River and Lake Orville, both of which are more than 400 miles away, and from the Colorado River. The engineering and political consequences of such projects are enormous, and demonstrate the increasingly delicate nature of our dependence on sources of water. Not only is there a scarcity in the quantity of water in many regions of the country, the quality of the nation's waters is subject to increasing threats.

Improperly sited, designed and operated landfills can accumulate significant quantities of dangerous chemicals which may leave the landfill area as runoff or leachate and enter streams that eventually empty into lakes, reservoirs or rivers which provide water for human use. Reported cases of contamination are legion and growing. In rural America, for example, there are numerous cases of surface water contamination which can be directly linked to landfills. A local creek near Shepardsville, Kentucky received the wastes from tens of thousands of discarded waste drums; in Byron, Illinois, local surface water and groundwater were contaminated with cyanide and a variety of toxic metals and metalloids from discarded containers of industrial waste; in Pichens, South Carolina, PCBs leaked from discarded electrical capacitors and transformers, contaminating a local watershed and the drinking water; in the Mississippi River a major fish kill in 1964 between Memphis and New Orleans was traced to contamination by indrin and other chlorinated hydrocarbons originated primarily from a solid waste dump in Memphis. In the Niagara Falls region, numerous landfills and dumps have contaminated groundwater and surface waters, including drinking water supplies, with toxic organic chemicals. The notorious Love Canal area is perhaps the best known example of how discarded industrial wastes have entered water supplies, leading to a major dislocation of residential communities.

Landfills can be made more "secure" to reduce the probability of contamination of surface and groundwater from disposed materials. They should be located in sites easily accessible to transportation, in areas of low population density and of low alternative land use value. To prevent leachate or runoff from accidental spills, landfills should be sited in low permeability, clay-rich soils in flat terrain, and should never be located over fractured bedrock. Landfills also should be located well above the high-water table, away from flood plains and wells which supply drinking water. Areas of low rainfall and high evaporation are preferred locations. The landfill pit should be excavated in clay, the subgrade compacted, smoothed, and fitted with an impermeable, inert liner of material such as asphalt, concrete, rubber, or plastic. The liner should be reinforced by covering it with a layer of compacted clay. A system of leachate drains should be constructed at the perimeter of the landfill base, adjacent to the liners, to collect for treatment any water leaking into the pit.

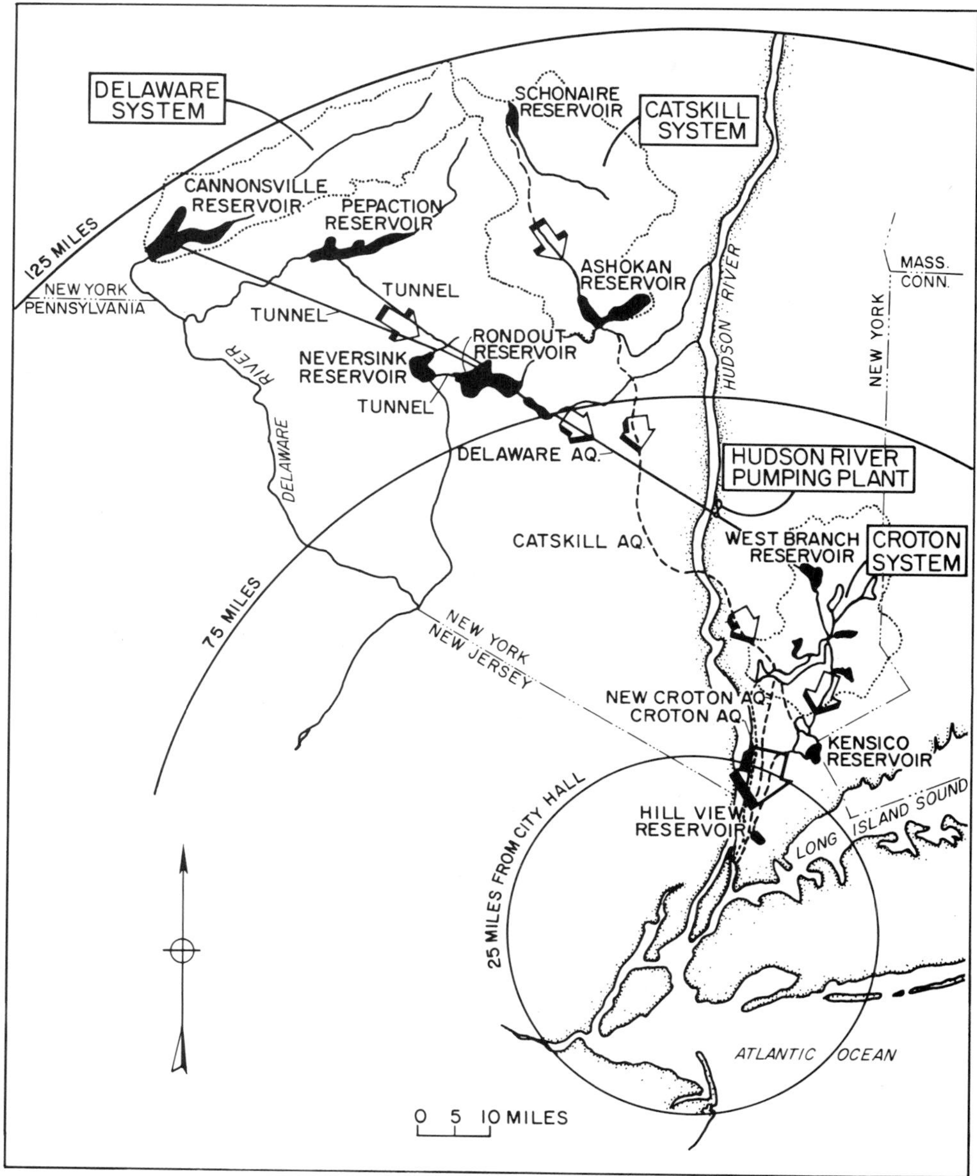


Figure 4-5 Sources of water for New York City.

Landfills for storage of toxic and hazardous wastes, secure landfills, require even more precautions. Secure landfills are divided by clay dikes into a series of separate compartments, each designated for specific waste streams. This permits the segregation of chemically incompatible materials--such as acid and cyanide wastes, which when mixed form highly toxic hydrogen cyanide--and facilitates possible future

Figure 4-6 Proposed water supply network for Los Angeles

recovery of any particular waste. When ready for closure, the landfill should be capped with an impermeable liner and a layer of clay. Each secure landfill should be monitored regularly to check for leaching at a network of wells which surround the site.

The most direct method of preventing pollution of surface waters in regions surrounding landfills is to prevent surface water from entering the landfill, and to ensure that all rain falling directly on the landfill and any surface water that does enter are collected, treated and released in a way that protects critical water resources. The terrain should be contoured to divert surface runoff around the landfill, and steps should be taken within the landfill to ensure that rainwater falling directly on it makes minimal contact with the compacted landfill cells.[2]

## 4-3 GROUNDWATER

Groundwater exists in massive quantities under most regions of the country. It has been estimated that there are over 150 million billion (i.e., trillion) gallons of groundwater at depths within one-half mile of the earth's surface.[3] This source would supply all of the needs of the entire world at the current use rate for the next 12.6 years without any replenishment, but because of the nature of groundwater containment and movement, remedial steps to remove many contaminants once it is polluted are complicated, expensive and often ineffective.

### Biological Contaminants

A primary measure of the quality of water is its freedom from pathogens, infectious bacteria and viruses. It is well known that over the centuries some of the most serious epidemics have been caused or sustained by polluted water supplies. Table 4-1 illustrates some of the many diseases known to be communicable through water supplies. There are many pathways by which viruses and bacteria from infected animals or humans may enter water supplies, but the primary source is excreta.

In measuring the quality of water, one standard is the number of coliform bacteria per unit volume of water. Since coliform bacteria are present in the fecal material of all warm blooded animals in enormous quantities (at the level of billions of bacteria per gram of material), the tracking of the coliform level provides a secondary method of estimating the presence of pathogens in water supplies.Current United States drinking water standards permit levels of coliform bacteria of 1 per 100 milliters of water. This measure of the quality of water is by no means universally accepted. The coliform standard assumes that a virus would be no more successful in surviving in the environment than coliform bacteria and that, indeed, if one observed only a single coliform bacterium in a 100 milliliter sample of water, the prospects of there being significant numbers of pathogens would be quite small and acceptable.

The role played by microorganisms in feces of warm blooded animals in causing human illness was first documented in 1854 by Dr. John Snow, an English physician. While investigating a cholera epidemic, Snow noted that one common link connecting those who became ill was their drinking water--it came from the same well. The issue was not simply geographic, where individuals in the area close to the well became ill. Some individuals who lived far from the well, who preferred the taste of its water and would often drink from it, also became ill. Later studies showed that the well was contaminated by leakage from a sewer.[4]

Since refuse brought to landfills comes from many different sources and inevitably contains components which carry pathogens, it is important to have some idea of the lifetimes of these pathogens in the environment. Such information is useful in estimating the probability that diseases might be picked up and subsequently re-transmitted to humans by animals or insects which visit a landfill.[5] Among the pathogens most dreaded by humans are those which establish permanent residency in the human body, where they find desirable temperatures and a favorable biological support system. As these pathogens multiply and invade sensitive areas of our bodies, we become ill. Almost every physiological function of our bodies can become compromised. The effects can range from headaches, stuffiness, dizziness, difficulty in breathing, rapid heartbeat, changes in blood chemistry, all the way to death. During the battles between the body's immune system and the invading pathogens, our body secretions contain residues of the war.

Excretions from the nose, lungs, mouth, and our urinary and intestinal tracts all may contain eggs, larvae, and adults of the organisms that caused the illness. To the extent that landfills receive even small amounts of materials contaminated with pathogens, the survival rates of these organisms in the landfill environment become relevant to human health. Table 4-2 summarizes the conditions, favorable and unfavorable, for the survival of human pathogens in the open environment.[6] Table 4-3 illustrates the survival rates of two types of intestinal pathogenic bacteria. Both *Salmonella typhi* and *Vibrio cholera* can survive for months in feces when frozen, and can survive for several days in food, water, and soil.

Viruses pose an even greater potential threat than bacteria to the public health because of their greater persistence and virulence. The available data clearly demonstrate the importance of using care in disposing of municipal refuse. Wastes from hospitals and nursing homes can be particularly troublesome.

### Chemical Contaminants

In addition to the possibility for biological contamination of water, there is the possibility of chemical contamination. Table 4-4 lists maximum permissible levels contained in the National Interim Drinking Water Regulations for several potential chemical contaminants.[7] No water sources designated for human consumption should exceed the concentrations given in this list. Other chemicals will almost certainly be added to the list.

In addition to chemicals known to cause illness in humans or in laboratory animals, many others give the water an objectionable taste and odor if present to an appreciable extent. Table 4-5 lists such chemicals and includes the maximum desirable concentrations of selected substances covered in the 1962 Public Health Service Drinking Water Standards.

### Aquifers

Next, we turn to the physical properties of groundwater systems, and the processes through which groundwater contamination occurs. A typical aquifer, the channel through which groundwater flows, is located 90 feet below the surface of the earth. It may be separated from the surface by a combination of sandy, silty and clayey soil strata, each characterized by a particular porosity and permeability which determine

**Table 4-1 Waterborne diseases transmitted by ingestion, grouped by types of etiological agent and ranked by likelihood of transmission [Ref. 1]**

| Disease | Comment |
|---|---|
| 1. Cholera | Initial wave of epidemic chlorea is waterborne. Secondary cases and endemic cases are by contact, food, and flies. |
| 2. Typhoid fever | Principal vehicles are water and food. Case distribution of waterborne outbreaks has a defined pattern in time and place. |
| 3. Bacillary dysentery | Fecal-oral transmission with water one transmitter. Direct contact, milk, food, and flies are other transmitters. Ample water for cleanliness facilitates prevention. |
| 4. Paratyphoid fever | Few outbreaks are waterborne. Other fecal-oral short circuits dominate. Ample water facilitates cleanliness. |
| 5. Tularemia | Overwhelmingly by handling infected animals and anthropoid bites. Drinking contaminated water infects man. |
| 6.Giardiasis | Carrier state up to 20 percent in United States. Outbreak in Rome, NY -- 4800 cases in 1974; 25 cases in Camas, Washington in 1976. |
| 7. Amebic dysentery | Epidemics, which are rare, are mainly waterborne. Endemic cases are by personal contact, food, and possibly flies. |
| 8. Infectious hepatitis | Epidemics are due to transmission by water, milk and food, including oysters and clams. |

**Table 4-2 Favorable and unfavorable environmental factors for the survival of human pathogens in the open environment [Ref. 1].**

| Favorable | Unfavorable | Comments on survival |
|---|---|---|
| Moisture | Drying | Drying is the most adverse condition. |
| Low temperature | High temperatures | Pathogens survive freezing; 60° C kills in 1 minute or less; 100° C is a certain killer. |
| pH range, 5-9 | pH below 5 and above 9 | Specific organisms have a narrower favorable range than 5-9; none are acidophlis or alkalophils. |
| Shade | Sunlight | The UV of sunlight and drying are killers |
| Freshwater | Saline water | Survival in freshwater is much larger than seawater. |
| Clean water | Polluted water | Competitive forms of life in polluted waters. |
| Sterile soil | Natural soil | Competitive forms of life in natural soil pathogens. |

the rate at which a leachate descends from the surface of the ground to the aquifer. The movement of a leachate from a source to the groundwater supply may take months or even years. Moreover, since the flow rate of the groundwater itself may be only a few feet per day, there can be an added and significant delay before the effect of a polluting source is detected at a distant site. Indeed, there are today many examples of groundwater contamination which have essentially "killed off" valuable aquifers and which had their origins in events decades ago at sites tens of miles away.

The significant lapse of time between the escape of pollutants into the environment and their entry into the groundwater system, and the exceedingly long residence times of some pollutants within groundwater systems, underscore the need for adopting a long range perspective in dealing with issues that affect the quality of groundwater. The siting, design and operation of landfills are among the most important issues to be considered.

**Table 4-3 Survival data on two intestinal pathogenic bacteria in the open environment [Ref. 1].**

| *Survival time* | *Salmonella typhi* | *Vibrio cholera* |
|---|---|---|
| 1. In feces | Min., 8 days<br>Max., 5 months<br>General, 30 days | 7-14 days unless frozen, then survives winter-long freezing |
| 2. In water | Sterile, 15-25 days<br>Tap, 4-7 days<br>Raw river, 1-4 days<br>Drainage canal, 2 days | 1-5 days<br>1-2 days<br>2-3 days<br>Rivers of India, 1/2-1 1/2 days |
| 3. On foods | Vegetables and fruits:<br>Min., 15 days<br>Max., 40 days<br>General, 20 days | On meat 7-14 days<br>On fish 3-4 days ambient temperature<br>On fish 10-12 days in icebox<br>Raw milk 1-1.5 days at 22-25°C |
| 4. On natural soil | 1 day<br>Max., 2 years in moist frozen.<br>General, under 100 days. | Moist tropical, 7 days<br>Russian winter, 4 months<br>Dry brick dust, 3 days max. |
| 5. With heating | In milk at 80°C,2 sec.<br>In milk at 62°C, 36-42 sec.<br>In milk at 60°C, 76-82 sec. | In water at 100°C, 0<br>In water at 80°C, 5 min.<br>In water at 56°C, 30 min.<br>In water at 40°C, 3 days |

By 1984 the United States Environmental Protection Agency had identified over 400 sites where groundwater contamination had reached such severity that specific remedial action will be required to prevent a direct threat to drinking water sources.[8]

The processes by which groundwater contamination occur are clear. Each year, a total of over 2 billion tons of solid wastes of all kinds are discarded throughout the United States. Moreover, of the total of 5,000 billion gallons of effluent treated by United States' industries each year, approximately one-third is placed in lagoons and oxidation ponds for treatment, leading to the leaching of an estimated 100 billion gallons each year into groundwater systems.[9] In addition, large quantities of liquid wastes are discharged illegally or accidentally.

**Table 4-4 Maximum permissible levels for inorganic chemicals in drinking water. [Ref. 1]**

| Element | Maximum permissible level (mg/l) | Reason for inclusion | Comment |
|---|---|---|---|
| Arsenic | 0.05 | Recognized poison; chronic effects, carcinogenic in some contacts, food intake contributory | Skin cancer high in areas of England with 12 mg/l in drinking water; $As^{3+}$ as $As^{4+}$ not essential or beneficial |
| Barium | 1.0 | Recognized toxic effects on heart, blood vessels, and nerves from accidental, experimental, and therapeutic ingestions | Water standard derived from occupational-exposure inhalation limit. Muscle stimulant |
| Cadmium | 0.01 | Acute poisoning in humans via foods, increased concn. in kidney and liver of rats on water with 0.1-10 mg/l | Individuals on water with average of 0.047 mg/l had no symptoms. Not essential or beneficial. Cigarette smoking a source |
| Chromium as hexavalent ion | 0.05 | Carcinogenic on inhalation; cumulative in rat tissue at level of 5 mg/l in drinking water; no toxic responses in rats for 1 yr at concn. of 0.45-25 mg/l | No observed effect on single exam. on family of 4 in 3 yr on water up to 1 mg/l. Not essential or beneficial |
| Lead | 0.05 | Recognized poison with daily intakes with food, water, air and inhaled tobacco smoke. Balance maintained at total intakes of about 0.3-0.4 mg/day | Intakes of 8-10 mg/l in water for several weeks are in the harmful range; poisoning reported from water varying from 0.04 to 1 or more mg/l; concn. as low as 0.1 mg/l is injurious to fish |

| | | | |
|---|---|---|---|
| Mercury | 0.002 | Recognized poision at work and in fish. Found in natural waters at less than 1 ug/l. Present is some foods from 10 to 70 ug/mg. Daily intake limit is 0.3 mg for 70 kg person | USFDA guideline for fish is 0.5 mg/kg. Fatal dose is from 20 mg to 3 g of salts. MCL for water provides for intakes from other sources. MCL rarely exceeded in drinking water |
| Nitrates as Nitrogen | 10 | Private well waters with nitrates from 15 to 250 mg/l caused methemoglobinemia in infants on milk formulated from such waters. Between 1945 and 1964 2,000 cases reported in U.S. and Germany with about 150 deaths. No cases from water under 10 mg/l | In only case of methemoglobinemia laid to public water supply, mother concentrated nitrates by long boiling. Older children and adults not effected. Intake of 2.2 mg of nitrate N per kg of body weight required to increase methemoglobin. Moldavian school children on 182 mg/l nitrate N showed a methemoglobin increase |
| Selenium | 0.01 | Recognized occupational poison and cause of livestock poisoning where Se exceeds 3-4 mg/kg of food intake; high in soils and crops in some localities in north central U.S. | Definite symptoms of poisoning via water very rarely have been identified; trace amounts believed essential for nutrition; mild poisoning in humans in high-Se areas observed |
| Silver | 0.05 | To limit additions of silver for disinfection; silver retention causes argyria, the blue-grey discoloration of skin, eyes, and mucous membranes | Human retention data are based on therapeutic use of silver compounds; drinking water limit calculated from l g total body burden which produces argyria |

In 1977 there were more than 18,500 municipal landfills in the United States. According to the Environmental Protection Agency, the majority were poorly sited and many were little more than open dumps without adequate safeguards to minimize the leaching of hazardous materials which collectively form over 50 million tons of the country's total wasteload.[10]

The preservation of the quality of groundwater supplies is critical. Over half of the United States' population relies on groundwater for drinking. People in certain areas of the country, for example the more than 3 million people in New York's Nassau and Suffolk counties on Long Island, rely totally on water drawn from aquifers for their needs. Groundwater supplies 25 percent of the total freshwater demand in the United States for all purposes.[11] It is a resource that should be protected vigorously.

### Leaching Process

One of the greatest environmental concerns associated with landfills is the potential for release of contaminants which may find their way into the groundwater. The potential for the slow, continuous release of undesirable materials onto the surface of the ground that may percolate through the soil and into the groundwater supply is of concern alone, but there is evidence that the situation may be even worse. The creation of certain landfills can cause a local rise in the level of the groundwater table itself. As a result, surface water can enter the groundwater channel following excavation of the landfill site, making groundwater pollution even more likely.[12]

There are two general ways to protect groundwater from landfill leachates. The first is to pump out groundwater in the vicinity of the landfill to lower the groundwater table, and thereby create an added filtration distance between the source of leachate and the groundwater. The second is to add as the foundation of the landfill a liner either of a plastic, such as polyvinyl chloride (PVC), or a thick layer of clay. The objective of a plastic liner is to trap the leachate. In the case of clay, the objective is to achieve an exceedingly slow filtration rate and to provide a large surface area on the clay particles to scavenge contaminants. In installing a plastic liner great care must be exercised to ensure that no punctures or tears occur during installation or in the initial subsequent disposal of garbage. This normally requires placing layers of sand above and below the liner for protection. The option of using a compacted clay liner avoids this problem, but containment of leachate by clay is not total. Clay impedes the flow of the leachate and provides opportunity for various biological and chemical reactions to occur within the clay to reduce or eliminate the toxicity of certain hazardous substances released from the landfill. In some circumstances, clay liners may be more effective than plastic liners.

In using liners to reduce the impact of landfills on groundwater, a problem results from the collection of rainwater by the large "tub" created by the containment devices. Because the water may contain highly concentrated leachates, it cannot be released directly into the environment. It must be collected and either recirculated through the landfill or treated before release; or be collected and disposed of as an industrial hazardous waste.

A typical landfill cross section is illustrated in Fig. 4-7. A bounded region on the surface overlies a clay stratum which, in turn, overlies a sandy aquifer underlain by bedrock.

**Table 4-5 Chemicals which give water objectionable taste and odor. Values tabulated are desirable maximum concentrations for substances related to consumer satisfaction, in accord with 1962 Public Health Service drinking water standards [Ref. 1].**

| Substance | Concentration mg/l | Reason for inclusion |
| --- | --- | --- |
| Alkyl benzene sulfonate | 0.5 | A nonbiodegradable component of synthetic detergents which persists through ground percolation and sewage- and water-treatment processes; foaming usual at 1 mg/l |
| Chloride | 250 | Proximate to salty-taste threshold; sudden increases result from sewage |
| Copper | 1 | Essential and beneficial for metabolism; taste threshold varies from 1 to 5 mg/l; limit prevents unpleasant taste |
| Carbon chloroform extract | 0.2 | CCE concentrations include part of the total organics in water, taste producers, toxicants, carcinogens, and wastes; water at 0.2 limit is already of poor quality |
| Iron | 0.3 | Essential and beneficial for metabolism, but water cannot meet the 7-35 mg daily requirement; proximate taste threshold, 2 mg/l; stains fixtures and white good at 1 mg/l; off flavors and colors in beverages, colloidal color in some water |
| Manganese | 0.05 | Staining of white goods by $MnO_2$ deposits, off flavors in bevarages; limit close to the attainable removal from most waters; probably an essential nutrient with 10 mg daily intake in food; toxic on inhalation |
| Phenols | 0.001 | Reaction products of phenolic compounds with chlorine cause objectionable tastes and odors |
| Sulfate | 250 | Laxative effect at 600-1000 mg/l when magnesium and sodium are the cations |
| Total dissolved solids | 500 | Taste and laxative effect the restraint; excess dissolved minerals in water result in poor brews of coffee |
| Zinc | 5 | Essential and beneficial in metabolism with daily intake of 10-15 mg; emetic action at 675-2280 mg/l; with zinc salts, milky at 30 mg/l and metallic taste at 40 mg/l. Limit is below taste threshold |

The processes governing the amounts of pollutants which leave the landfill and may appear later in the groundwater can be thought of as occurring in two distinct phases. In the first phase, soil near the discharge attenuates the leachate by adsorption and absorption. In the second, further attenuation of the leachate occurs in the aquifer itself by dilution and by adsorption and absorption onto particulate matter.[13]

The effectiveness of landfills in containing leachate from garbage and trash depends greatly upon the design of the landfill, the type of soils within the region of the site, the proximity to aquifers and, of course, the composition of the material deposited in the landfill. Depending upon the chemical composition of the leachate, various reactions occur in the underlying stratum to attenuate the leachate. However, once the residual leachate reaches the aquifer, dilution is the only effective mechanism of attenuation. There are many examples which demonstrate that it is often insufficient to protect the quality of groundwater.

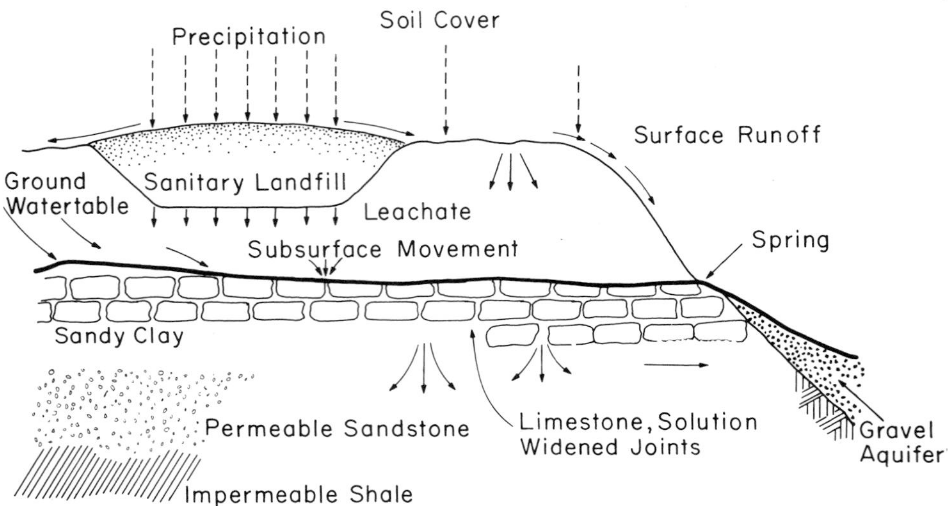


Figure 4-7 Cross-section of landfill near aquifer

## 4-4 PARTICULATE MATTER

The atmosphere we breathe contains not only the mixture of elements we know as pure air, but other chemicals, as well as actual particles. Such particles range in size from about 0.00004 inches (0.1 micron)--about the size of large molecules--to 0.02 inches (500 microns)--about the diameter of a walrus whisker. The larger particles, defined as aerosols, are generated as stoker fly ash, industrial dusts and pollens. The smaller particulates include smoke, fumes from combustion and paint pigment. The particle size range for aerosols is given in Table 4-6.

Aerosols are generated through a variety of processes, natural and anthropogenic. In many sections of the country particles on dirt roads made airborne by passing traffic are the primary contributors to wind-blown dust and soil. Another major contributor to atmospheric particulate matter is construction activities. Table 4-7 indicates the percent of the aerosols in selected urban areas attributable to different sources. According to these data, the contribution to the total burden from refuse incineration facilities is less than that from coal burning, motor vehicle emissions and cement manufacturers.

Approximately 10 million tons of particulate matter enter the atmosphere daily from natural sources and roughly 700,000 tons from anthropogenic sources. Of special concern is the fraction, roughly 1 percent, of the particulates that carry metals. Small amounts of lead, cadmium, chromium, arsenic, nickel, and mercury are present in the atmosphere from a variety of sources. It is known that these metals can be quite toxic when inhaled in significant quantity. As discussed in Chapter 6, lead, which can enter the waste stream through a variety of mechanisms, can cause physical handicaps, mental retardation and even death. Cadmium can enter the waste stream through burning of automobile tires and zinc smelting activities.

The degree of danger posed by the inhalation of particulates depends not only on the composition but also on the sizes of the particles. Particles larger than 0.0006 inches (15 microns) normally are expelled by breathing. Smaller particles not only pose the risk of being retained within the lungs, but may penetrate deep within the lung tissue and thereby transfer their metal contaminants to the bloodstream.[14] Moreover, smaller particles are most likely to carry lead, cadmium and other contaminants because their surface to volume ratios are greater, providing more opportunity for adsorption and absorption.

## 4-5 ODORS

One of the first reactions provoked by the mention of the word garbage is concern about odor. Smells produced by decaying organic material are objectionable. In the home they are a reminder to take the garbage out. Odors also can be objectionable while garbage is awaiting collection for transport from the home to a disposal site or transfer station, and at the disposal site itself. In this section we address odors and how they relate to garbage as it decays at landfill sites and is burned in resource recovery facilities.

Since odor is defined not by how much of a particular contaminant is in a unit volume of air, but by how the human nose reacts to the substance, odor can be quantified only statistically. Studies of smell almost always focus on measures of the intensity of odors, their threshold for detection, their acceptability, and how they are related to other odors.[15]

Intensity is a measure of the strength of an odor. Is it faint or overwhelming? An odor's detection threshold is the concentration of a substance (in air) permitting it to be detected by the human nose. Acceptability is a measure of whether the odor is regarded positively or negatively. Is it foul or enticing?

The intensity of an odor can be established only on a statistical basis using actual human responses. One method for quantifying odor intensity is to use an odor panel, usually 6 to 12 people, who independently record their reactions to an odor. For studying intensities of odors, measurement units are created so that 1 odor unit corresponds to the amount of the substance contained in 1 cubic foot of air which

**Table 4-6. Particle size ranges for aerosols [Ref. 14].**

| Substance | Approximate Range of Particle Mean Diameter (microns) | | |
|---|---|---|---|
| Rain Drops | 500 | --- | 5,000 |
| Natural Mist (water vapor) | 60 | --- | 500 |
| Natural Fog and Clouds (water vapor) | 2 | --- | 60 |
| Stoker Fly Ash | 10 | --- | 800 |
| Pulverized Coal Fly Ash | 1 | --- | 50 |
| Foundry Dusts | 1 | --- | 1,000 |
| Cement Dusts | 3 | --- | 100 |
| Metallurgical Dust | 0.5 | --- | 100 |
| Pollens | 10 | --- | 100 |
| Ground Talc | 0.5 | --- | 50 |
| Bacteria | 0.3 | --- | 35 |
| Plant Spores | 10 | --- | 35 |
| Sulfur Trioxide Mist | 0.3 | --- | 3 |
| Insecticide Dusts | 0.5 | --- | 10 |
| Pigments (paint) | 0.1 | --- | 5 |
| Ammonium Chloride Fume | 0.1 | --- | 3 |
| Alkali Fume 5 | 0.1 | --- | |
| Oil Smoke | 0.1 | --- | 1.0 |
| Metallurgical Fume | 0.01 | --- | 2.2 |
| Resin Smoke | 0.01 | --- | 1.0 |
| Tobacco Smoke | 0.01 | --- | 1.0 |
| Normal Impurities in Quiet Air | 0.01 | --- | 1.0 |
| Carbon Black | 0.01 | --- | 0.3 |
| Colloidal Silica | 0.02 | --- | 0.05 |
| Zinc Oxide Fume | 0.01 | --- | 0.5 |
| Magnesium Oxide Fume | 0.01 | --- | 0.5 |
| Sea Salt Nuclei | 0.03 | --- | 0.5 |
| Combustion Nuclei | 0.01 | --- | 0.1 |
| Virus and Protein | 0.003 | --- | 0.05 |
| Gas Molecules (diameter) | 0.0001 | --- | 0.0006 |
| Dust Damaging to the Lung (silicosis) | 0.5 | --- | 5 |
| Human Hair | 35 | --- | 200 |
| Red Blood Cells (adults) | 7.5 | --- | |

**Table 4-7 Air pollutant emissions in United States in 1982. Amounts given in millions of metric tons.**

| Source | Carbon monoxide | Sulfur oxides | Organic compounds | Particulates | Nitrogen oxides |
|---|---|---|---|---|---|
| Total emissions | 73.6 | 21.4 | 18.2 | 7.5 | 20.2 |
| Transportation: total | 53.3 | 10.9 | 6.1 | 1.3 | 9.7 |
| Road vehicles | 46.3 | 0.5 | 4.8 | 1.1 | 7.8 |
| Total fuel combustion (stationary sources) | 6.6 | 17.4 | 2.0 | 2.4 | 9.6 |
| Electric utilities | 0.3 | 14.3 | <0.05 | 1.0 | 6.2 |
| Industrial processes | 4.8 | 3.1 | 7.1 | 2.4 | 0.6 |
| Solid waste disposal | 2.1 | <0.05 | 0.6 | 0.4 | 0.1 |
| Misc. uncontrollable | 6.8 | <0.05 | 2.4 | 1.0 | 0.2 |
| PERCENT OF TOTAL EMISSIONS | | | | | |
| Transportation | 72.4 | 4.2 | 33.5 | 17.3 | 48.0 |
| Fuel combustion | 9.0 | 81.3 | 11.0 | 32.0 | 47.5 |
| Industrial processes | 6.5 | 14.5 | 39.0 | 32.0 | 3.0 |
| Solid waste disposal | 2.9 | 0 | 3.3 | 5.3 | 0.5 |
| Misc. uncontrollable | 9.2 | 0 | 13.2 | 13.3 | 1.0 |

Source: Statistical Abstract of the United States (1985), 105th Ed. (Table No. 346)

can be detected by half of the panel members, but not by the other half.[16] Different individuals have different sensitivities to different odors and can reach varying conclusions regarding the strengths, as well as the acceptability, of the same odor. This is why a statistical approach must be used.

The nose is a remarkable detector of certain compounds. It has a sensitivity to concentrations as low as three parts of a contaminant in a billion parts of air--a concentration of 3 parts per billion. This sensitivity is like demanding that the human eye look at a crowd consisting of every human being in the United States and pick out the presence of a single individual.

The chart shown in Table 4-8 indicates the odor thresholds for a variety of compounds. The left hand column shows the concentrations of the substances which are detected by half the members of an odor panel, but not the other half. The right-hand column shows the minimum concentrations detectable by the entire panel. With such great sensitivites it is apparent that care must be exercised to ensure that even those chemicals not thought to be harmful to health are regulated to avoid degrading the aesthetic quality of life.[17]

The decay of garbage produces a variety of gases responsible for the objectionable and hazardous odors at landfills. Just as steps must be taken to contain surface runoff and to protect groundwater supplies from leachates generated within the landfill, care also must be taken to see that lateral gas seepage does not occur. Depending on the geological structure underlying a landfill, it is possible for low resistance paths to exist that make it easy for gas to pass great distances from landfills causing odors, unhealthy conditions, and even the potential for explosions at points well beyond the landfill boundaries. In areas where the surficial sedimentary deposits have a high permeability, methane gas can migrate over distances of several miles creating the potential for explosions and other health hazards. Long Island, New York is an example of such a situation.

## 4-6 VECTORS

Rodents, insects and birds are attracted to landfills, particularly to those which are poorly operated. The supply of food is continual and the variety impressive. In a large metropolitan area the continuously replenished supply of nourishment for rodents, birds and insects provided by poorly maintained landfills poses a serious public health problem for nearby residents, and for public health in general. Rodents and other pests are carriers -- vectors -- for many pathogenic organisms. If there are any lessons to be learned from history, one is that every attempt must be made to control the populations of species known to be vectors for disease, and to control access of those species to waste products.

Any situation which gives insects and rodents access to infected material has the potential for creating dangerous human health conditions. Studies in 1932 found that 44 percent of the flies trapped in the rooms of patients with typhoid fever carried typhoid fever bacteria and that active bacteria were present in the fly's intestinal tract for six days after removal of the flies from the original environment.[18] Access of vectors to wastes in landfills from hospitals and to diapers from residences and nursing homes can be particularly hazardous.

Insect populations, including flies, mosquitos, cockroaches, ticks and mites flourish near landfills. Although sanitary landfilling technology can greatly reduce the potential for problems from insects and rodents, it can never totally eliminate them. The ineffectiveness of means of controlling rodents and other pests is one of the clear disadvantages of landfilling compared with other types of solid waste disposal.

**Table 4-8 Typical odor threshold values [Ref. 14].**

| Compound (Increasing Threshold) | Threshold Value, ppm ( by volume) 50% Response | 100% Response |
|---|---|---|
| Amine, Trimethyl | 0.00021 | 0.00021 |
| Ethyl Acrylate | 0.0001 | 0.00047 |
| Hydrogen Sulfide Gas | 0.00021 | 0.00047 |
| Butyric Acid | 0.00047 | 0.001 |
| Ethyl Mercaptan | 0.00047 | 0.001 |
| p-Cresol | 0.00047 | 0.001 |
| Dimethyl Sulfide | 0.001 | 0.001 |
| Sulfur Dichloride | 0.001 | 0.001 |
| Benzyl Sulfide | 0.0021 | 0.0021 |
| Methyl Mercaptan | 0.001 | 0.0021 |
| Diphenyl Sulfide | 0.0021 | 0.0047 |
| Nitrobenzene | 0.0047 | 0.0047 |
| Pyridine | 0.01 | 0.021 |
| Amine, Monomethyl | 0.021 | 0.021 |
| Phosphine | 0.021 | 0.021 |
| Amine, Dimethyl | 0.021 | 0.047 |
| Benzyl Choloride | 0.01 | 0.047 |
| Bromine | 0.047 | 0.047 |
| Chloral | 0.047 | 0.047 |
| Phenol | 0.021 | 0.047 |
| Diphenyl Ether | 0.1 | 0.1 |
| Styrene | 0.047 | 0.1 |
| Acetaldehyde | 0.21 | 0.21 |
| Acrolein | 0.1 | 0.21 |
| Carbon Disulfide | 0.1 | 0.21 |
| Methyl Methacrylate | 0.21 | 0.21 |
| Monochlorebenzene | 0.21 | 0.21 |
| Chlorine | 0.314 | 0.314 |
| Allyl Chloride | 0.21 | 0.47 |
| Methyl Isobutyl Ketone | 0.47 | 0.47 |
| p-Xylene | 0.47 | 0.47 |
| Sulfur Dioxide | 0.47 | 0.47 |
| Acetic Acid | 0.21 | 1.0 |
| Aniline | 1.0 | 1.0 |
| Phosgene | 0.47 | 1.0 |
| Formaldehyde | 1.0 | 1.0 |
| Toluene Diisocyanate | 0.21 | 2.14 |
| Toluene (from petroleum) | 2.14 | 2.14 |
| Benzene | 2.14 | 2.14 |
| Toluene (from coke) | 2.14 | 4.68 |
| Perchloroethylene | 4.68 | 4.68 |
| Ethanol (synthetic) | 4.68 | 10.0 |
| Methyl Ethyl Ketone | 4.68 | 10.0 |
| Hydrochloric Acid Gas | 10.0 | 10.0 |
| Carbon Tetrachloride | 10.0 | 21.4 |
| Acrylonitrate | 21.4 | 21.4 |
| Trichloroethylene | 21.4 | 21.4 |
| Ammonia | 21.4 | 46.8 |
| Dimethylacetamide | 21.4 | 46.8 |
| Dimethylformamide | 21.4 | 100.0 |
| Acetone | 46.8 | 100.0 |
| Carbon Tetrachloride | 46.8 | 100.0 |
| Methanol | 100.0 | 100.0 |
| Methylene Chloride | 214.0 | 214.0 |

## 4-7 BIRDS, LANDFILLS AND AIRPORTS

Because competition for space is intense in many large metropolitan areas and because landfills and airports both require relatively large areas they are often located near each other. This has been particularly true for large coastal cities. Landfills and airports both attract things that fly -- sometimes into each other. The bird population near Kennedy International Airport in New York City, as well as at many other major airports in the world, is influenced by nearby landfills. A count at one of the dumps revealed over 10,000 gulls feeding at one time. These landfills are like fast food restaurants for birds, and for gulls in particular. The ready availability of food creates leisure time which the birds often choose to spend in the general vicinity of the landfill (and the airport), often soaring at heights which pose a hazard to aircraft.[19]

Gulls often soar at altitudes of up to 3,000 feet over landfills, taking advantage of both the currents of warm air which rise above the landfills and, presumably, the attractive odors. Any bird traffic at such heights within a few miles of a busy airport can be dangerous.

At Kennedy Airport, 42 percent of the 130 bird strikes reported by pilots from 1973 to 1981 occurred at take-off and 58 percent during landings. The take-off period is most critical because increased power is needed and an unexpected loss of an engine can cause disaster.[20] In Germany, between 1970 and 1976 Lufthansa Airlines reported 2,247 bird strikes.

The October, 1960 crash of a turbo-prop plane at Boston's Logan Airport was believed to have been the direct result of the plane flying through a flock of starlings. Sixty-two passengers lost their lives in this tragic event.

As airport traffic becomes more intense, greater care must be taken in siting landfills. Succeeding generations of birds are becoming increasingly accommodated to their urban surroundings. The availability of food, without expending effort, offers birds enticements which outweigh the disturbance they experience as a result of noise and activity at nearby airports.

Studies in Australia have suggested that when possible landfilling near major airports might best be restricted to nighttime, with the deposits being covered before morning to reduce attractiveness to birds.[21]

## 4-8 VEHICLE EMISSIONS

Once garbage has been delivered to landfills, it must be transported to the proper area within the landfill boundary, spread into a thin layer, compacted, and covered with soil. These operations cause large landfills to be areas of high vehicular activity, with bulldozers, large refuse hauling trucks, refueling trucks, water sprinkling trucks, dirt transport trucks continuously in motion throughout the day and night. The amount of diesel fuel required, and the resulting vehicle emissions, are not insignificant. In the case of the Fresh Kills landfill in New York, in 1985 more than 3,280 gallons of diesel fuel were required daily (1.2 million gallons per year) in handling the disposal of the more than 20,000 tons of refuse received by that facility each day. Fuel consumption is predicted to increase to 1.5 million gallons per year in 1986.

## 4-9 GAS POLLUTION

Depending on the moisture content of the garbage in a landfill, either aerobic or anaerobic decay will dominate the decomposition process. Anaerobic reactions lead to the production of methane gas and care must be taken to see that concentrations do not build up to levels that can lead to explosions. To avoid this problem, the cells containing buried garbage must be vented to release the gas. Under certain conditions the release of methane from landfills is so copious that several communities are capturing it for use as an energy source.

By 1980, 23 landfill gas (LFG) systems had been developed in California.[22] In 1983, 26 such systems were in operation on Long Island, New York. These account for the bulk of LFG activities in the United States.[23] Fresh Kills landfill on Staten Island, New York has the world's largest methane recovery facility. The plant which is capable of processing almost 10 million cubic feet per day of raw landfill gas, began operating in 1983.[24] Gas is withdrawn from over 100 wells, 65 to 75 feet deep, on 400 acres of the Fresh Kills landfill. The total yield is almost 1.3 billion cubic feet of gas per year.[25] The gas is transported in vacuum by an underground collection system to a plant where it is processed to remove trace elements, carbon dioxide and moisture. The final product is nearly pure methane, with a heating capacity equivalent to that of natural gas--1,000 BTU per cubic foot. The purified landfill gas is mixed with natural gas and distributed to customers by Brooklyn Union's west shore facility.[26]

## 4-10 A NOVEL APPROACH TO LANDFILLING

Several municipalities in flat-lying areas of the country have chosen another landfilling strategy: to pile the trash into enormous heaps. These isolated mountainettes of municipal waste have been dubbed Mt. Trashmores and have been centerpieces of planned recreational areas. Virginia Beach's Mt. Trashmore is typical of this new breed of man-made mountains.

In 1966 Virginia Beach faced a problem. The existing 50-acre dump was a semi-open trench with only a short lifespan remaining. The high water table, 6 to 8 feet, made pit excavation impractical. The cost of solid waste disposal was already high, $75,000 per year, and was increasing. Roland Dorer, Director of the Virginia State Health Department, proposed the idea of extending the existing sanitary landfill upward to increase its capacity and therefore its lifespan. Constructing a hill of solid waste would not pose the health hazards associated with pit excavation. And so in 1966 the City began construction of a 68 foot high hill of garbage and trash. Over 640,000 tons of refuse were dumped on the site over a four and one-half year period. With a tipping fee of approximately $1.97 per ton, the total cost to construct Mt. Trashmore's was $1,115,095. The high density achieved by daily compaction of alternating layers of 18 inches of trash and 6 inches of soil created what has proven to be a stable configuration.

Through this project low-value land adjacent to the Norfolk-Virginia Beach Expressway was converted into a 68 foot high hill with a soap-box derby ramp, foothills, freshwater lakes, skateboard ramps, parking areas, concession stands, and playground and picnic areas (Fig. 4-8). There is no evidence of adverse

Figure 4-8 Mt. Trashmore recreation area; example of creative final use of landfill.

environmental effects from Virginia Beach's Mt. Trashmore. Seven seepage points for evacuation of decomposition gases have eliminated potential build-up concentrations of methane to explosive levels. Odor is not a problem, because of the special care employed in providing daily cover.

During the construction stages there was some public opposition because of odors, flies, gulls, and gases. Once these were alleviated, Mt. Trashmore became a source of local pride. Today, Virginia Beach's Mt. Trashmore is a valuable recreation area. it was so well received that Virginia Beach decided to build another and larger Mt. Trashmore. Construction was begun in 1971. When the project is completed in 2015, a 573 acre landfill will have been converted into a recreational area which will include two artificial lakes and a 140 foot high Mt. Trashmore. Perhaps it should be called Mt. Trashmost.

## 4-11 SPECIALIZED LANDFILLS

### Ashfills

When refuse is incinerated, approximately 20 percent of the mass remains as ash.[27] The majority of existing and planned incinerator facilities use, or plan to use, landfills for disposal of ash residue. Because incinerator ash, particularly fly ash, contains elevated levels of many heavy metals, and of dioxins and furans, ashfills are an environmental concern (see Chapters 6 and 7 for further discussion).

### Balefills and Shredfills

Baling or shredding refuse decreases its volume by allowing the waste to be compacted to a higher density. The need for cover at balefills and shredfills may be greatly reduced since rodents and insects cannot exploit these fills as easily as conventional landfills. Blowing paper is also minimized. An important aspect of balefills is the fact that the leachate generation rate is less than at conventional landfills.

### Landfills on Distinct Sites

Sanitary landfills are often used in conjunction with reclamation of land lost to other activities. Abandoned clay pits, quarries and strip mines are amenable to this waste disposal method. Strip mines are well suited to landfilling since the underlying stratum is usually impermeable clay or shale. Siting is not a problem since strip mines are not located in populated or environmentally sensitive and valuable areas. Canyons and ravines are also used as landfills. Sanitary landfills should never be constructed in wetlands. Dumping refuse in these water-saturated environments greatly increases the potential for leachate contamination.

**4-12 FINAL USE**

Once a landfill has reached capacity it can be converted to other uses. The characteristics of the solid wastes which went into it must be considered carefully when deciding which use is most appropriate. Landfill density will vary according to the waste composition, climate, topography and adequacy of the landfill design. Settlement of landfills ranges from 2 to 40 percent, with an average value of 20 percent of the initial height.[28] Building on completed landfills is greatly discouraged at least during the first twenty years after the fill is closed. Most completed landfills are used for recreation or as open space. Even for these uses, maintenance is required.

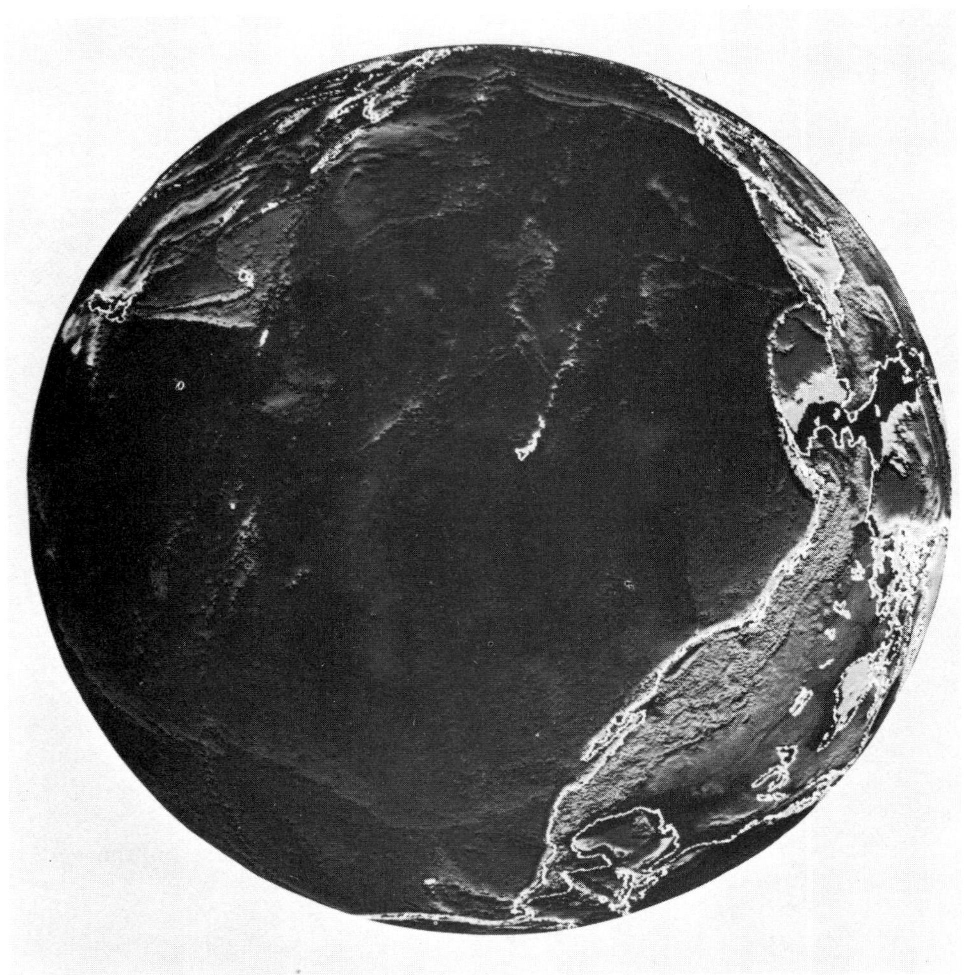

Figure 5-1 Satellite photograph of earth.

# CHAPTER 5

## ENVIRONMENTAL IMPACT OF OCEAN DUMPING

### 5-1 INTRODUCTION

The surface of the earth is dominated by the oceans ( Fig. 5-1) which cover over 70 percent of the earth's surface. Their enormity has long been a source of awe and inspiration. On a geological time scale, only very recently was it discovered that the extent of the oceans is finite and that the earth is round. Before the age of discovery, many believed that the oceans extended to infinity. Even up to the present, society's views of the vastness of the oceans have affected the ways in which we treat oceans as a receptacle for our waste products.

The oceans are the eventual repository, the ultimate sink, of almost all pollutants. Industrial, municipal and other wastes discharged into rivers and streams eventually end up in the oceans. Some cities accelerate this process by pumping their wastes directly into the ocean through pipes or by barging them--millions of tons of sewage sludge, dredged materials and other wastes--moderate distances offshore for direct dumping into the ocean. Moreover, a large fraction of the pollutants discharged into the atmosphere eventually precipitate into the ocean.

We now realize that the oceans do not have infinite capacity for indiscriminate dumping of wastes without incurring an environmental debt that must be paid by future generations. The oceans connect environmentally essentially all corners of the world. DDT is an example frequently cited of how a waste discharged at one point can affect marine life great distances away. DDT was found in Antarctic penguins even though there had been no use of DDT within thousands of miles of the Antarctic.[1]

The oceans are the principal source of food for at least 10 percent of the world's population. Through evaporation, the oceans provide the precipitation that falls on land. Clearly the future quality of all human life is intimately connected to the quality of the oceans. But the connections of society to the quality of the land and the atmosphere are no less important. Because of this all three media--air, land, and water--should be carefully assessed in selecting disposal strategies. The oceans do have a capacity to receive certain amounts of society's wastes without suffering persistent, undesirable change. This threshold varies with the category of waste, and with the times, methods and locations of disposal.

Using historical data, it is difficult to separate the activities and effects of ocean disposal of municipal solid wastes from the disposal of other solid wastes such as sewage sludge, dredged material and industrial wastes. At present, no municipality in the United States dumps its garbage and trash into the ocean.

In this chapter we briefly review the history of dumping of garbage and trash into the ocean by the United States, and examine some of the effects of municipal solid waste disposal on the ocean and its living resources.

### History

The use of the ocean as a waste receptacle dates back hundreds, probably thousands of years. Until recently, most waste disposal occurred in estuaries that discharge into the ocean rather than directly into the ocean itself. The 1902 Encyclopedia Britanica described ocean disposal as a "clean method" of waste disposal apparently because human activities did not appear to be adversely affected by it, and because nutrients in the waste were assumed to increase production of fish, shellfish and seaweed. Ocean disposal has other advantages. It is easy, convenient and economical. But disposal in the marine environment has caused problems.

In December, 1675 Governor Edmond Andros, the second English governor of the colony of New York, forbade any person to "cast any dung, dirt, refuse of ye city or anything to fill up ye harbor or among ye neighbors under the penalty of forty shillings".[2] In spite of this New York City continued to dump its waste products into the sea, but moved the locus of its activities farther downstream. For the past 50 years, it has concentrated its ocean dumping in the New York Bight at several sites 12 miles off the entrance to New York Harbor. As a result of a 1985 United States Environmental Protection Agency directive, the dumping of sewage sludge must be moved to a site 106 miles off-shore.

Because of public pressure and local ordinances, the dumping of raw garbage and trash into the ocean by the United States began to decline about 1950, and today no municipality dumps its municipal solid wastes into the ocean. The United States Ocean Dumping Act of 1972 and the International Dumping at Sea Act of 1974 both banned ocean dumping of many of the components of garbage and trash. The latter Act is particularly important because it ensures that possible future decisions by the United States to implement ocean dumping will require approval at the international level. Substances specifically banned by this Act from ocean dumping include metals from printers' inks, plastics, wood, organohalogens, radioactive wastes, and biological and chemical warfare agents.

Most municipal solid waste dumping activities and their effects on the ocean and its living resources have been poorly documented. While no United States municipalities now dump their garbage and trash into the ocean, large amounts of these materials still find their way into the ocean. Some is blown off barges which transfer trash from municipalities to coastal landfills; some is blown off piers and beaches; and some is discharged through combined storm drain/sewer systems. But most is intentionally thrown into the ocean from vessels of all kinds and sizes.

### Boats As A Source of Garbage and Trash

Virtually all the garbage and trash from the world's commercial fleet is thrown directly into the ocean. If any treatment is involved, it is only to compact it. And this is done on only a small percentage of vessels. It is difficult to estimate accurately the volume of material discarded from vessels, but it is not trivial. In the world fleet there are 25,580 ships with combined crews that total more than 625,000.[3]

Those who work and live on ships produce less garbage and trash on a per capita basis than the general public. Newspapers are not routinely delivered on the high seas, and space limitations aboard ship dictate conservative practices in purchasing and in packaging. The space problem is most acute on submarines. Submarines compact their garbage and trash because there is little room onboard. In situations which call for avoiding detection, the garbage and trash is placed in weighted bags which sink to the sea floor. At other times much of it floats to the surface. In addition to the commercial and military fleets, there is the recreational fleet. In the United States alone the number of registered pleasure vessels in the marine zone totals more than 4.5 million.[4] While it has not been documented, we suspect that much of the garbage of this fleet, perhaps most of it, and some of its trash is thrown directly into the coastal ocean. Compared with the area and volume of the world ocean, the input of garbage and trash from ships is small, but it is not insignificant, at least not locally in coastal waters.

The United States' National Academy of Sciences estimated that the total amount of trash discarded into the ocean from all kinds of craft (passenger, merchant, military, commercial fishing and recreational boats) was about 7.0 million tons each year.[5] This did not include garbage. The estimated composition of this litter is shown in Table 5-1.

**Table 5-1. Composition of trash discarded into the world ocean from vessels (Data from National Academy of Sciences, Ref. 5).**

| Material | Percent by Mass |
|---|---|
| Paper | 63.0 % |
| Metal | 16.6 % |
| Cloth | 9.6 % |
| Glass | 9.6 % |
| Plastics | 0.7 % |
| Rubber | 0.5 % |
| | 100.0 % |

Plastics account for only a small fraction of the total amount of trash thrown into the ocean, but they are persistent, ubiquitous, and regarded by some as insidious (see Fig. 5-2). Plastic litter is distributed throughout the world ocean. Most is polystyrene. An average of 12,000 pieces and 2.1 pounds of plastics per square

...l mile were reported in the Sargasso Sea in 1971.[6] The ocean surface con-...tion of plastics in surface waters ranged from 1 to 10 ounces per square nautical mile.[7] Similar concentrations have been reported in other areas of the open ocean.[8,9] Besides plastic, the most abundant kinds of ocean surface litter are wood, metal (cans, floats, etc.) and glass. Some of the trash thrown into the ocean ends up along its shore.

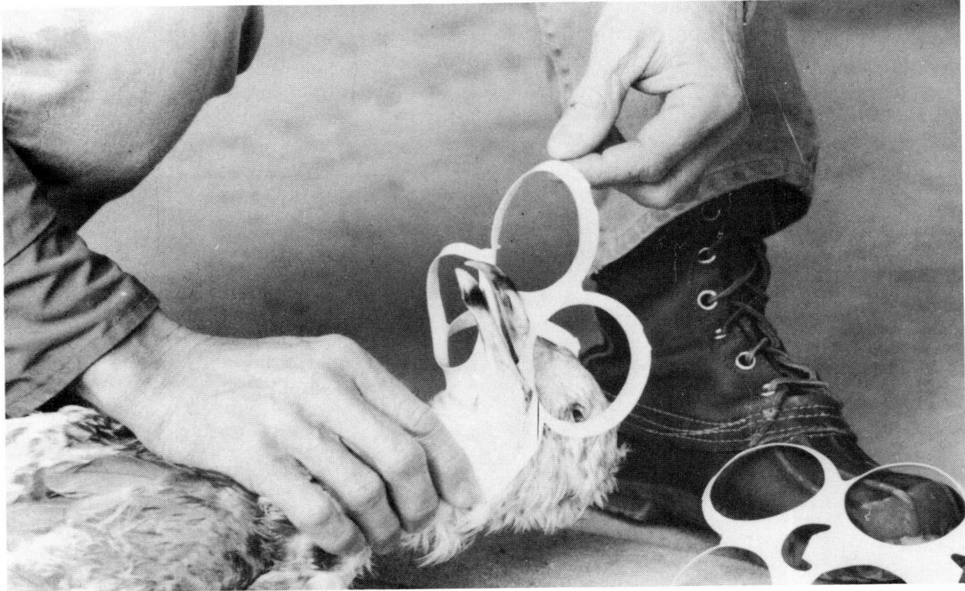

Figure 5-2 Birds entrapped in plastic 6-pack beverage container fixtures

---

The kinds and concentrations of litter on beaches has been quantified in only a few areas. Scotland and England have well over 1600 items of litter per mile of beach.[10,11] The kind of beach litter and the size of individual items vary widely. In the Aleutian Islands 55-gallon oil drums and large pieces of lumber are numerous. In other areas the most common litter consists of polyethylene pellets about 0.2 inch in diameter. These pellets, called "nibs" by the industry, are used extensively in packaging. They have been reported in concentrations that average over 10,000 per yard of beach front near Auckland, New Zealand.[12] On selected Long Island beaches, concentrations ranging up to thousands of spherules (0.1 - 0.8 inch diameter) per cubic inch of sand were observed.[13]

In 1977, along the shoreline of Scotland, plastics made up 32 percent of the total litter volume, followed by wood (27 percent), glass (12 percent), paper and cardboard (12 percent), metal (8 percent), hemp and rope (5 percent), and other materials (4 percent).[14] The source of litter was believed to be refuse discarded by commercial ships. Changes in the relative abundances of the constituents compared with those in the original wastes discarded, reflect differences in persistence and "floatability". In Narragansett Bay (Rhode Island), the major source of beach refuse is the recreational boater.[15] On Long Island's south shore beaches, New York City's combined storm drain/sewer system is a major contributor of litter and, under certain conditions, can be the dominant source.

## 5-2 EFFECTS OF TRASH ON THE MARINE ENVIRONMENT

The dumping of raw, unprocessed, unsorted garbage and trash is objectionable because of the high percentage of floatables which can be dispersed over wide areas, fouling beaches and posing a threat to marine life, particularly marine birds. The impacts of trash--the inorganic fraction of municipal solid waste--are not the same as the impacts of garbage--the organic fraction. The impacts of trash are biological, aesthetic and economic. Trash can harm birds, fish, mammals, and vessels. It also is aesthetically objectionable, and as a result, can adversely affect the economy by discouraging recreation. In the water, and particularly on the bottom, trash can foul and tear fish nets.

### Biota

Several species of birds ingest litter. Plastic spherules about 0.04 to 0.2 inches in diameter were noted in the food pellets routinely regurgitated by terns and gulls on Great Gull Island in eastern Long Island Sound.[16] The gizzards of Leach's petrel nesting near Newfoundland contained small polyethylene nibs.[17] Petrels generally feed in surface waters far at sea. Polyethylene nibs were observed in the crops of dead black and mallard ducks on Southern Long Island (New York). No obvious harmful effects from ingesting the plastic particles were noted in cursory examination, but the investigators expressed concern and urged more careful study.

Litter may have had a detrimental effect on the puffins of Scotland and Norway.[18,19] Puffins ingested elastic threads (like rubber bands) which drift down river to the sea. Numerous dead puffins with apparently clogged guts were observed. The threads appeared to have caused obstruction of the puffins' alimentary canals. It is uncertain whether this contributed to the marked decline of the British puffin population.

As with birds, there is very little information on the effects of litter on fish, but there is the suggestion of possible deleterious effects. Members of eight species out of fourteen in Long Island Sound contained polystyrene spherules in their guts.[20]

Of the white perch and silversides larvae examined, one-third contained plastic spherules. Many of the larvae were only 0.2 inches long and 0.04 inches across and yet contained spherules 0.02 inches in diameter. The potential for intestinal blockage is apparent.

In the Severn estuary (United Kingdom) flounders less than two years old contained polystyrene spherules. Some larvae only 0.8 to 2 inches long had as many as thirty spherules each in their intestines.[21] However, the supply of spherules to the Severn estuary may have been eliminated, or at least reduced significantly, since the incidence in flounder dropped from 20.7 percent of those examined in the Spring of 1974 to none in the Spring of 1975.[22]

In laboratory studies, scientists observed ingestion of spherules by five fish species common to the southern New England coastal zone,[23] but no plastics were observed in the gut contents of over 500 larvae and juvenile fishes (22 species) collected off the east coast of the United States in 1972.[24]

It is apparent that little is known about the frequency of occurrence of plastic litter in fish intestines and the effects on the organisms. Fish in estuaries and nearshore waters may have a higher incidence of litter ingestion than those further offshore. Marine mammals also are affected by litter. It has been reported that as many as 8,000 fur seals became entangled in drifting polypropylene netting in the north Pacific in 1977.[25] Many of these seals perished from the restricting net materials.

Litter is not without it benefits in the marine environment. Larger items can provide shelter for small fish.[26] And it is well known that the largest of debris, such as sunken ships, provide habitats for large fish.[27,28]

#### Fishermen's Equipment and Ships

The costs to fishermen from entanglement and tearing of trawl nets on bottom debris is great.[29] The fouling of ships' screws and water intakes by floating debris also causes considerable economic loss. In 1969, the total loss to shipping from debris in those American ports which handled one-third of all shipping tonnage was estimated at $17.4 million.[30]

#### Beaches

In the summer of 1976 long stretches of Long Island's ocean beaches were closed because of stranding of objectionable floatable materials related to garbage and trash and to sewage. A significant fraction of these materials was attributed to material blown off New York City garbage barges, to pier fires in New York Harbor, and to the water-borne transport of materials floating in New York Harbor.[31] Curiously, little of it could be linked to the intentional dumping of sewage sludge in the New York Bight. Beach cleanup costs for this one event were put at $100,000, and the loss to beach-related industries was $15 to 25 million.[32] The annual cost of removing trash and litter from beaches is high. Clean-up costs for less than 2 miles of public beaches in Bermuda is more than $100,000 per year. Much of the Bermuda litter consists of tar balls and plastic debris.

### The Sea Floor

Very little is known about the amount and composition of trash on the sea floor. In 1975 in the Gulf of Alaska, a variety of kinds of trash were observed: plastic materials such as garbage bags, bait wrappers and cargo binding materials, and lost fishing gear.[33,34] On the continental shelf off the Netherlands, large numbers of man-made articles such as packing materials, bottles and metal drums were reported.[35]

## 5-3 ORGANIC WASTES FROM SEAFOOD PROCESSING PLANTS AND CANNERIES

The ocean also has received organic wastes from seafood processing plants and canneries. Fruit and vegetable cannery wastes consist of ground fruit pits, skins and pieces of fruits and vegetables. Such inputs to the ocean were concentrated in a few locations and the amounts varied dramatically with season. Seafood processing wastes consist of the waste water and the remains of fish and shellfish (offal).

Between 1960 and 1972, 246,000 tons of fruit and vegetable wastes from canneries in the San Francisco Bay area were dumped into the ocean. An average of 22,000 tons, mostly overripe peaches, pears and skins, were dumped each year at a site 20 miles offshore. The wastes, were mixed with water, pumped overboard and discharged at a depth of about 15 feet. Fouling of beaches was common because most of the components were floatable.[36] Dumping was terminated in 1972 because of the cost of extensive monitoring which was then required.

## 5-4 GARBAGE AND TRASH DUMPING OFF THE PACIFIC COAST

Between 1931 and 1971 there were three active garbage disposal sites on the Pacific coast, all off California.[37] The site off San Francisco Bay primarily received cannery wastes, discussed previously. The other sites were located 20 miles off Newport beach near Santa Catalina Island and 20 miles off Monterey. Both were on the lower slope of the southern San Pedro Basins. Little information is available on the composition or amount of material dumped off Monterey. It is known that the Monterey and Newport Beach dumpsites received 26,000 tons of wastes in 1968 and 21,000 tons in 1971, with about 85 percent of the total being dredged materials.[38]

About 158,000 tons of garbage and trash were dumped off Newport Beach between 1931 and 1970. Roughly 73 percent of it was dumped by the United States Navy between 1944 and 1970;[39] the remainder was dumped by commercial vessels. Unprocessed materials--domestic garbage, paper, cartons, cans, bottles, metal parts from ship repairs, ropes, rags and old clothes--were dumped at the surface of the ocean at the rate of about 600 tons per year between 1931 and 1970. Problems from dumping activities included beach-fouling with floatables and short dumps caused intentionally or by poor navigation. The State of California suggested that garbage and trash be reduced to a pulp by maceration prior to dumping to reduce problems with floatables and that a navigation method more accurate than dead-reckoning be used to prevent missdumps.

In 1958 the United States Navy dumped garbage and trash from Navy and commercial vessels, along with other material, in surface waters off California, using scows holding 25 to 136 tons.[40] The high percentage of plastics and other floatables caused aesthetic problems and fouling of beaches and may have been a hazard to birds, mammals, and fish. Detailed effects of dumping off California were never seriously studied.

## 5-5 SITE SELECTION AND TRANSPORTATION COSTS

When considering the feasibility of using the ocean as disposal site for garbage and trash, or the residues of burned garbage and trash, one must evaluate the economics of such disposal relative to other disposal options such as landfilling and incineration in resource recovery facilities. Transportation costs represent a major fraction of the total costs of most refuse disposal systems. From a direct, short-term economic standpoint, refuse should be dumped close to shore. But health, social, and ecological factors loom larger as the distance of a dumpsite from shore decreases.

In addition to economic advantages, there are other advantages of dumping at a nearshore rather than at a deep-sea site.[41] These include: (1) missdumps caused by storms are less frequent, (2) monitoring of wastes on the bottom is easier (3) the dumpsite may be used as a fish-attracter, and (4) benthic organisms in nearshore areas are more capable of dealing with stresses caused by dumping than organisms in the deep-sea which are accustomed to a more stable environment. Deep-sea dumpsites, on the other hand, offer the advantages that: (l) the probability is reduced that refuse will be displaced from the bottom because of decreased surface wave and current activity and because the wastes are compacted by high hydrostatic pressure; and (2) interference of dumping operations with recreational activities, commercial boating and the fishing industry are decreased.[42,43] The question as to whether deep-sea or nearshore dumpsites should be used for garbage and trash is unresolved, and will require further comparative studies if ocean dumping is permitted in the future as a viable municipal solid waste disposal method.

Detailed economic appraisals have been made to compare the transportation costs of ocean disposal with costs of landfilling (the only other ultimate disposal option).[44,45,46] These appraisals are presented here as examples only. The ultimate economic assessment of ocean dumping compared to landfilling is strongly dependent on local factors and the nature of the assumptions underlying the economic appraisal. In a study conducted for Westchester County (New York)[47] disposal of solid wastes by barge-haul and disposal of baled refuse at-sea was found to be technically feasible and less costly than rail-hauling combined with sanitary landfilling or incineration. The study did not recommend at-sea disposal, however, because of the difficulty and uncertainty in determining the ecological effects of ocean disposal. Devanny and colleagues have compared the economics of land-based and sea-based municipal solid waste disposal on the basis of direct costs of dumping.[48] The conclusion of this study, based on an economic analysis for a city such as New York, was that if one considers only the direct costs of disposal, ocean dumping is an economically feasible option. When considering the costs of ocean dumping one must consider, however, the degree of processing of trash required prior to dumping. This is addressed in the following sections.

## 5-6 THE OCEAN DISPOSAL OPTION

Municipal solid wastes can be disposed of in the ocean in a variety of forms: loose refuse, compacted and baled refuse, pulp, loose (unstabilized) incinerator ash, and incinerator ash stabilized into solid blocks. Each option has different economic and environmental costs and benefits. In the sections that follow, we consider each of these options.

### Loose refuse

In this option, trash is delivered to dockside, loaded onto barges, transported to the dumpsite, and dumped. This practice has been used in the past, but it has obvious drawbacks. Trash is approximately 50 percent paper which along with other floatables, will float or remain suspended in the surface layers of the ocean. Floatables can be dispersed rapidly over a large area. Since paper is easily decomposed, this may be only a short-lived problem, but if the frequency of dumping is high, the problem will be a persistent one.

### Dumping of compacted bales

In this option, trash is delivered to a baling facility, shredded and compacted into bales and strapped. Baling decreases the problem of floatables and bales can be used to construct fishing reefs. In this way, the waste products may become a resource.

Loder and colleagues baled garbage with a modified hydraulic concrete testing machine and placed the bales in the ocean to determine whether or not they would act as suitable substrates for various animal populations.[49] Glass and metal were added to increase the density of bales to make them sink. After compaction, bales were wrapped in polypropylene mesh and strapped. Two tubes were inserted into each bale to permit sampling of interstitial waters. Some bales contained food; others did not. The dumping site had a sandy seafloor and good circulation.

Mobile organisms increased in abundance on and around the bales, but no change was detected in the animals which lived in the bottom.[50] Certain fouling organisms (hydroids) settled directly on the waste material; others (Spirorbids) settled only on the inert mesh baling and strapping.[51] Successful colonization of the bales may depend on the nature of the strapping material, the types of organisms found in the area of the dump, and textural differences between the baled garbage and the baling material. Crevices may be selected by organisms in preference to smooth surfaces.

The bales maintained their physical integrity over the year they were studied, although they lost the metal baling clips because of corrosion. Within a short distance of the bales the concentrations of potentially toxic, dissolved degradation products formed within bales dropped to ambient levels, and posed no threat to marine life in the immediate area.

Pratt and colleagues conducted a laboratory study of the biological effects of ocean disposal of baled solid wastes.[52] Their objectives were (1) to define the effects of dissolved leachates from baled garbage and trash on the physiology and behavior of marine organisms; (2) to characterize the loss of solid materials from baled garbage and its subsequent dispersal; and (3) to describe the interactions of fouling and motile organisms with the waste blocks.

The solid wastes used in their study were composed of food, tin cans, aluminum, plastic, glass, and paper. The paper component was 72 percent of the total, which is higher than that usually found in typical domestic wastes. The food content of the bales was either 9.8 percent or 2.2 percent and was composed of dog food, rice, and sugar.

Because of decomposition of the organic matter (primarily paper) in the bales, solutes such as hydrogen sulfide, phosphate and ammonium built up to very high levels within the bales.[53] However, because of the high density of bales, toxic substances seldom escaped. Toxic metals were essentially trapped within the bales because of metal sulfide precipitation. Unfortunately, high bale density often prevented escape of gases like methane, hydrogen sulfide and nitrogen. Because of gas buildup many bales actually became buoyant and floated to the top of the experimental enclosures. These data indicate if baling is used to process wastes for ocean disposal, steps must be taken to achieve a compaction density to reduce toxic solute escape and to keep gas buildup at levels below which bales become buoyant.

The studies showed that baled refuse forms a favorable habitat for organisms such as ciliated protozoa, oligochaete worms, nematodes, hydrozoans, tunicates and harpacticoid copepods. The exact species composition of colonizing communities of the bales is dependent on the surface substrate, local current speed and the presence or absence of bacterial mat formation. Long-term studies showed that baled refuse had little effect on the material community of animals living in the sediments in the dumping area. This indicates that baled refuse may have no adverse effects and may even positively influence the ecology of areas in which dumping occurs. Careful monitoring of baled refuse will undoubtedly be necessary in possible future applications of this method to ensure that local areas of the ocean are not adversely affected.[54]

### Unstabilized Ash Disposal

Ash generated from waste incineration usually requires disposal, even though recycling reduces the final amount of material. Since many municipalities are already committed to resource recovery, or are considering it as a method of reducing solid waste, questions concerning incinerator ash disposal must be addressed. The oceans represent one possible receptacle for incinerator ash, but a great deal of additional research will be necessary to assess potential environmental, public health and economic impacts.

Incinerator residues include potentially toxic substances such as heavy metals, furans and dioxins. The extent to which these substances are leached from ash placed in the marine environment and enter the marine food web will be important in determining the appropriateness of the ocean as a disposal site. Very little research has been done on disposal of incinerator ash in the ocean.

When dumped in unstabilized form, incinerator ash may pose many of the same problems as unstabilized raw garbage and trash. Ash may adversely affect the biota at the dumpsite. In laboratory and field studies, quahog, winter flounder, shrimp, adult lobsters, and millet were not significantly affected by floating incinerator ash, whereas menhaden, lobster larvae and sea scallops experienced substantial increases in mortality because of ash ingestion.[55,56] Heavy metal toxicity was thought to be the major cause of animal mortality, although in some cases, organisms showed a decrease in metal accumulation rates upon exposure to incinerator ash.

A relatively large proportion of incinerator ash is floatable and may foul beaches. When incinerator ash was deposited at a site two miles off Rhode Island, movement of materials was sufficiently great that fouling of beaches would have occurred with heavy dumping.[57] Ash movement was attributable largely to wave action during storms. When ash was dumped at a site 18 miles offshore, in deeper, calmer waters, very little ash transport occurred. Dispersion of materials was so limited that no beach fouling would have occurred, even with very heavy dumping. This study indicates that deep-sea sites probably would prove to be less objectionable than near-shore sites for disposal of unstabilized incinerator ash.

The unstabilized ash that sinks to the bottom may adversely affect benthic communities. While no studies have focused specifically on the effects of incinerator ash residues on benthic community structure, one study did investigate the effects of coal fly ash. Because the physical properties of coal fly ash and incinerator fly ash are similar, some extrapolation of the results of the coal ash study are probably valid, although direct analyses would be preferable.

Pulverized coal fly ash coagulates in the ocean to form agglomerates that may be a yard or more in length.[58] When the aggregates settle to the bottom they cause mortality by burial, and there is little or no tendency for recolonization by species which live on the bottom. When ash particles are resuspended from the bottom and subsequently ingested, aggregates form in the intestines of organisms and inhibit nutrient uptake, causing mortality. Although future studies will be necessary to assess the potential impacts of incinerator ash on benthic populations, existing information suggests that if ash is dumped in an area, a large portion of the marine food web--the benthos--may be entirely removed from the local dumping area.[59] This and other factors suggest that ocean dumping of stabilized ash probably is preferable to disposal of loose ash.

### Stabilized Ash Disposal

Stabilization of incinerator ash is any procedure that reduces leaching of toxic substances and floating or resuspension of small ash particles. Work is just beginning to determine the optimum procedure for stabilization of incinerator ash for placement in the marine environment.[60] This work is based largely on an earlier five-year study of coal fly ash stabilization by the State University of New York at Stony Brook's Marine Sciences Research Center.[61] Certain differences between coal fly ash and incinerator ash (particularly the higher levels of metals and dioxin in the latter) prevent complete extrapolation of the results of coal ash stabilization studies to incinerator ash stabilization, although some insights may be gained.

In the Stony Brook study, coal fly ash and scrubber sludge (mainly calcium sulfite hexahydrate) were mixed in varying proportions and formed into blocks weighing about 60 pounds each. The objective was to determine if the blocks could be used to construct artificial reefs that would attract marine life without adverse environmental impacts.

In 1980, a reef was constructed of 15,000 fly ash-scrubber sludge blocks at a site five miles off the south shore of Long Island in about 70 ft. of water (Figs. 5-3, 5-4). The reef was monitored for more than five years and compared to a control reef constructed of demolition rubble. Results showed that the fly ash-scrubber sludge reef enhanced local primary and secondary productivity of the area. Growth of animals on the experimental reef was very similar to that on the control reef. There was little tendency for the experimental blocks to lose their compressive strength or to corrode over long periods of time. Faunal analyses indicated that there was very little change in the ecology of the surrounding bottom area. Various long-term exposure tests showed little or no tendency for increased animal mortality or toxic metal uptake by fish in the presence of the blocks, indicating that the toxic potential of the fly ash had been essentially eliminated by the stabilization procedure.[62]

Stony Brook's reef studies of coal wastes indicate that stabilization of incinerator ash and marine disposal may be a practical strategy. Because reefs attract marine life, disposal of stabilized ash to form reefs may actually have beneficial impacts on the marine environment rather than the adverse impacts associated with disposal of unstabilized ash. Future studies will be necessary to determine whether or not stabilization of incinerator ash can be accomplished using technologies that will compete with other disposal methods in terms of economics, aesthetics and environmental impacts.

Figure 5-3 Sketch of the Stony Brook Marine Science Research Center's artificial reef constructed from blocks of stabilized coal wastes, including racks of test bricks.

Figure 5-4(a) During first month on the sea floor (70 feet deep), the reef blocks are yet to be colonized. Particulates and mucous strings which can be seen in the photograph are naturally occurring in the bottom seawater at the project site. (b)After about 13 months on the sea floor, an abundance of marine life inhabits the reef blocks. Cunner feed about the reef blocks which are overgrown by epifauna, principally a branching bryozoan (*bugula turrita*).

## 5-7 CREATIVE USES OF GARBAGE AND TRASH: ISLANDS AND BREAKWATERS

In 1974 New York City proposed construction of a resource recovery and park (RECAP) island. The island was to be constructed on the ocean floor by erecting a dike, filling it with compacted refuse, and capping it. A resource recovery facility was to have been located on the island with an 11,000 ton per day capacity which would produce 7,700 tons per day of refuse-derived fuel. Electricity generated from this fuel would have met 5 percent of New York City's electrical demand. The project was never implemented, primarily because the costs and risks were considered to be too high. Other potential uses of garbage for construction were considered on Oahu, Hawaii. An evaluation was made to determine if garbage could be used to construct a breakwater, a surfing beach, and an extension of the Honolulu Airport facility.[63,64] None of these ideas was implemented because of high costs and uncertain technologies. Projects such as these may, however, become feasible alternatives in the future, as the costs of disposal by other means become higher, the available options become more limited, and as new technologies evolve.

## 5-8 CONCLUSIONS

The ocean evokes sympathetic responses from many and emotions run high against use of the ocean as a receptacle for society's waste products. Still, no matter how conscientious we are about source reduction and recycling, municipal solid waste products will remain which will require disposal somewhere in our environment. Our choices are limited: the atmosphere, the land, the ocean. The ocean is one of our alternative waste disposal sites. It should be evaluated carefully and critically in the context of the other alternatives. The advantages and disadvantages of each alternative need to be assessed on a regional basis for different kinds of wastes. Only then can we make the best--most appropriate--choice, rather than having the selection made by default with the wastes going to the least regulated (protected) component of our environment. The ocean does have the capacity to accept certain amounts of waste before suffering unacceptable (unreasonable) degradation. That capacity varies significantly for different kinds of wastes. It also varies significantly from place to place within the ocean, and with time at a given location. Under the proper conditions, and in proper forms, certain "waste products" can be used to enhance the ocean and coastal environments. They can be used to stimulate primary productivity, and to create artificial fishing reefs which increase--at least locally--secondary productivity. Some stabilized waste products can also be used for construction in the ocean, along its margins and on land.

The public health, environmental, and economic impacts of the ocean alternative for disposal of municipal solid wastes have not been rigorously evaluated. They should be; for disposal of raw garbage and trash, for baled garbage and trash and for disposal of resource recovery ash in both unstabilized and stabilized forms.

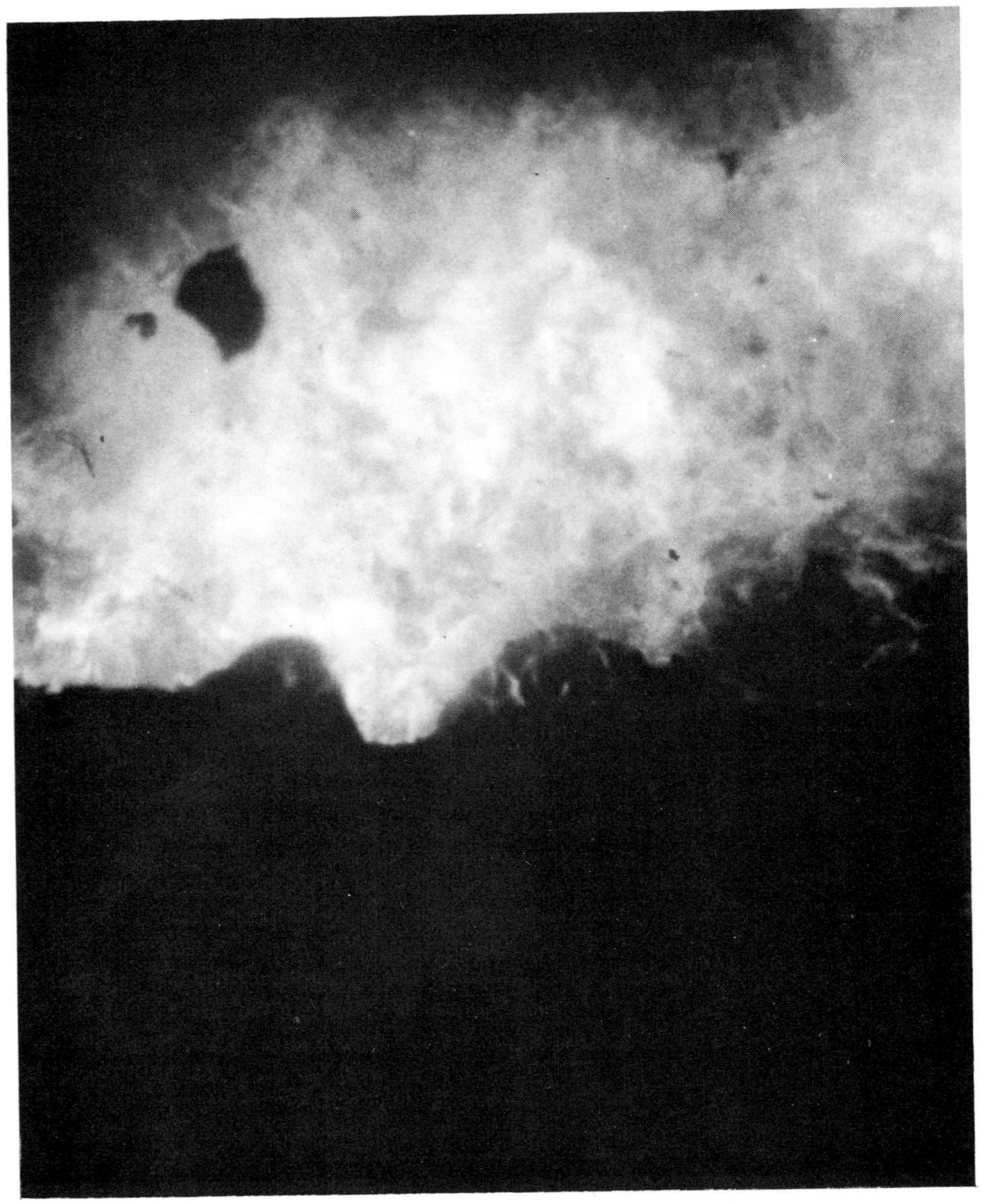

Figure 6-1 Garbage being incinerated in resource recovery facility.

# CHAPTER 6

## RESOURCE RECOVERY INCINERATION TECHNOLOGIES

### 6-1 INTRODUCTION

Approximately 350 resource recovery incinerator facilities are operating in over 15 countries in the world.[1] The shared feature of these facilities is that garbage is placed in huge furnaces and incinerated at high temperatures, usually from 1,650 to 1,850 degreees Fahrenheit. In these installations arrays of tubes embedded in the walls of the furnace carry water to recover the heat generated from the burning waste. Heat transferred to the water from the burning trash and garbage causes the water to boil and to produce steam. Normally the steam is used to turn turbines which produce electricity. In other cases, the steam is transported to nearby industries or homes for direct heating.

The residue (ash) from the incineration process normally contains significant amounts of metals, including iron and aluminum. These materials can be separated from the ash and sold to dealers for recycling. This recovery of resources--energy and useful materials either before or after burning--gives rise to the term "resource recovery". This concept clearly will become more and more relevant in dealing with garbage disposal in the future.

Ideally, a resource recovery facility accepts garbage for incineration and captures a large fraction of the energy from the combustion process for use as electricity or heat and, either before or after incineration, extracts all materials that can be recycled. The incineration temperature should be sufficiently high to break down any chemicals that would be harmful if released into the atmosphere, but not so high that heavy metals, which are themselves harmful, are vaporized and released. Nor should the temperature be so high that the reliable long-term mechanical operation of the facility is compromised. Using technologies developed in Europe starting in the 1920s, there are now a number of resource recovery facilities operating throughout the world which come close to these ideals.

Communities considering solid waste disposal options must take into account the relative economics of mass incineration and landfilling, as well as the environmental and public health impacts these two options may have on the region. For landfills the primary issues are the possibility of groundwater contamination and the cost of land in highly populated regions of the country. For resource recovery facilities, the primary issues have to do with the extent to which emissions may pose health hazards, and the capital costs of the construction project.

In this chapter we review the design and operation of resource recovery facilities, the nature of the emissions, possible impacts of the operation of these facilities on surface water and groundwater, a comparison of the emissions that would result from a standard fossil-fuelled power plant producing the same amount of energy as a resource recovery plant, and strategies contemplated for dealing with disposal of the residual ash. In Chapter 7 a comparative summary is made of the economic and environmental considerations applicable to the landfilling and resource recovery disposal options.

## 6-2 OPERATION OF RESOURCE RECOVERY FACILITY

Normally a resource recovery facility is designed to handle the garbage and trash of a county, bi-county, or regional area. To take a specific case of an operating facility, we will focus on the Pinellas County resource recovery plant serving the greater St. Peterburg, Florida, area. A photograph of this plant is shown in Fig. 6-2. A schematic drawing of a typical modern resource recovery facility is shown in Fig. 6-3. The Pinellas County Refuse-to-Energy Facility, which was put into operation in 1983, handles 2,100 tons of garbage per day and serves a population of about one million. When present upgrades of the Pinellas County facility are complete, it will provide 75 megawatts of electric power to the Florida Power Corporation. This electric power will serve the equivalent of 56,000 homes.

Garbage trucks pick up in residential and commercial areas and transport their roughly 10 ton loads to the resource recovery facility. On entering the facility, each truck is weighed and the driver questioned about the contents of the load to insure that no highly flammable items are included, and to anticipate the presence of any large items (e.g., refrigerators) that might best be removed before incineration. The driver then takes the load to the tipping floor of the facility and backs up to a large pit. In the Pinellas facility, this pit is 50 feet wide by 250 feet long and has a depth ranging from 35 to 65 feet. It is capable of holding up to three days worth of garbage and trash deliveries, or 6,000 tons of garbage; the equivalent of approximately 600 standard truck loads.

Material is removed from the pit by cranes controlled by operators situated in cabins high above the pit. A crane's huge sector-clawed mechanical hands can lift over one ton of garbage in a single bite and deposit it in a chute leading to the furnace. The garbage is gravity fed into the furnace where it is hydraulically rammed onto the moving grate system.

As fresh garbage and trash enter the stoker the material is pushed on top of burning garbage from deeper in the furnace which has been returned close to the entrance chute by action of the reciprocating grate--a characteristic of the patented Martin process described later in this chapter. In this way waste entering the furnace for the first time is quickly dehydrated and combusted.

The primary solid residue of the combustion process collects at the bottom of the grate and is called bottom-ash. It falls onto conveyor belts which carry it to other locations within the facility where it is automatically sorted by size of the particles in the residue. After incineration, particles of a certain size will fall onto one conveyor belt, the remnants of a refrigerator onto another. Bottom residue within a given size range is then passed near a magnet which removes the ferrous pieces for sale to iron dealers. The bulk of the other metal is aluminum from beverage cans. As discussed in Chapter 8, this residue also can be collected and sold for recycling.

Figure 6-2 Pinellas County, Florida refuse-to-energy facility.

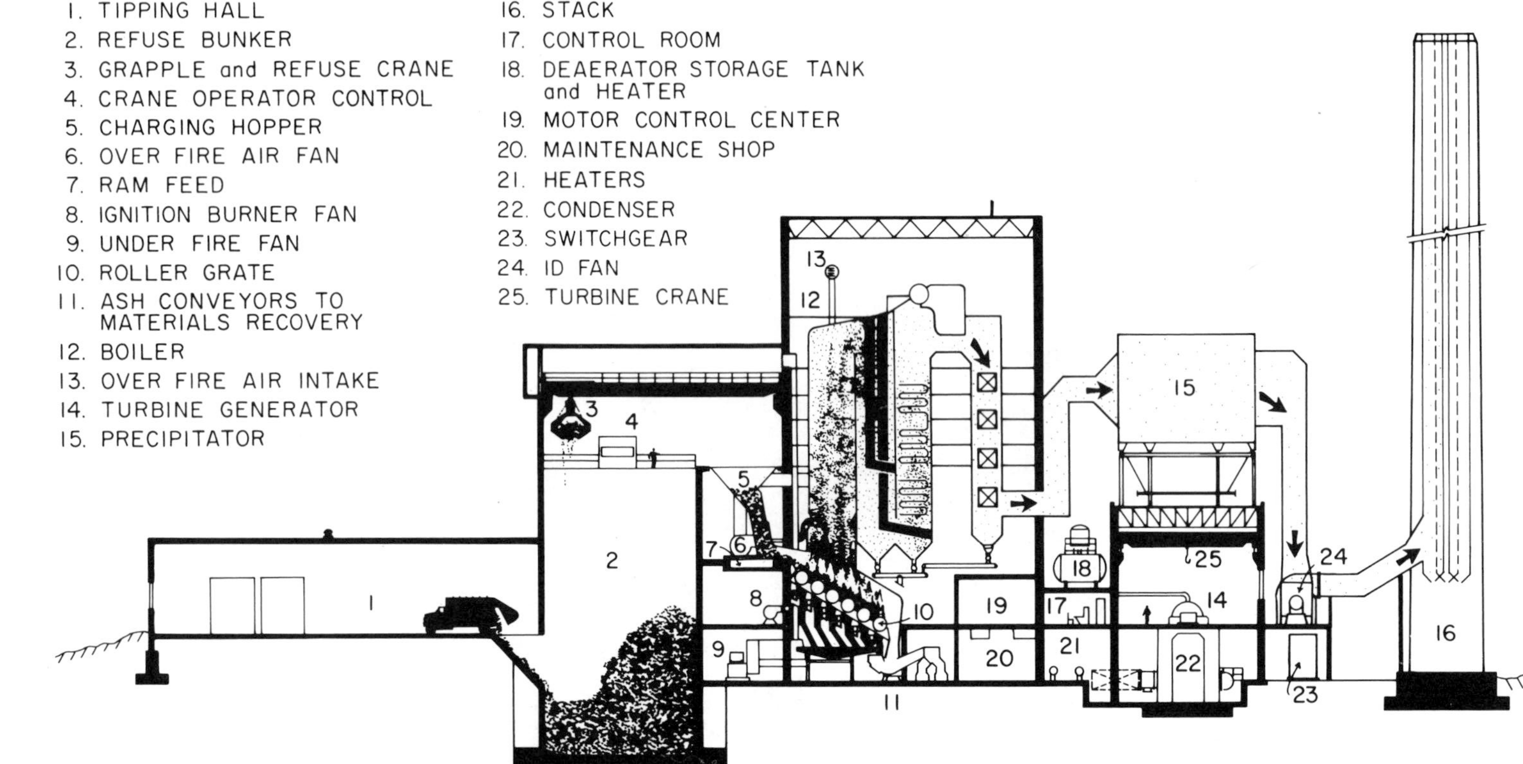


Figure 6-3 Schematic of typical modern resource recovery facility.

In addition to bottom ash, fly ash--made up of particles suspended in the effluent gas--is generated by the combustion process. Normally fly ash is collected in devices called electrostatic precipitators. In these large devices, placed in the path of the gaseous flow of the emissions en route to the smoke stack, fly ash is given an electrical charge and is attracted to plates in the precipitator, where it is trapped. It accumulates on the plates and is removed periodically. Usually it is mixed with bottom ash and disposed of. Other devices called baghouses or fabric filters are also being used today in some resource recovery facilities. These devices function somewhat like a nylon stocking placed over the exhaust of a household clothes dryer.

Fly ash is very fine-grained, not unlike soot from fireplaces. For every ton of garbage burned, approximately one-quarter ton ends up as some form of ash.[2] Fly ash accounts for about 10 to 15 percent of the total ash residue; the remaining 85 to 90 percent is bottom ash. The ash from most resource recovery facilities is buried in landfills, or used as landfill cover. Because ash contains metals, dioxins, furans and other contaminants, there is concern that burying the ash in landfills may hasten the entry of these contaminants into the environment and the water supply. Although past United States Environmental Protection Agency toxicity tests have concluded that resource recovery ash is not to be classified as a hazardous waste, it is clear that continued research on ash disposal options and its creative uses is desirable. Various options for the utilization of ash, including use as aggregate in concrete blocks, are discussed below.

Figure 6-4 summarizes the products that result from processing in a resource recovery facility the one ton of solid waste generated annually by an active United States citizen. The products are roughly 500 kilowatt hours of electric power, 500 pounds of solid residue (125 pounds of ferrous scrap material, 325 pounds of material suitable for use in road building or other construction, and 50 pounds of material that must be landfilled), and emissions of various gaseous materials discussed later in the Chapter.

Appendix 4 contains a selected list of resource recovery facilities in operation, under construction or planned in the United States.

## 6-3 COMBUSTION PARAMETERS

The efficiency with which municipal solid waste is incinerated in a furnace depends upon a variety of combustion parameters, including the furnace temperature, the amount of air injected into the furnace, the degree of turbulence, the uniformity of the burning bed and the time period each element of garbage and trash is exposed to high temperatures. An improperly designed and operated furnace can simultaneously have regions where air is insufficient to sustain burning and nearby regions which receive so much air that torch-like hot spots exist. It is important that such conditions be avoided for a variety of reasons. On the one hand, incomplete combustion leads to the emission of excess quantities of a variety of compounds ranging from the components of black smoke to potentially harmful hydrocarbons. On the other hand, excessive temperatures in isolated hot spots can lead to grate damage and slag formation. The capability for avoiding such conditions is one of the features which distinguishes good and bad furnace designs.

A well-designed furnace should quickly dry the waste and bring it to combustion. It should vigorously mix air uniformly throughout the waste during combustion and ensure that each element of garbage and trash and its combustion products are held at a temperature above 1,800° F for at least one second. Some data indicate that, at least within a controlled laboratory environment, these combustion conditions destroy more than 99.9 percent (by weight) of many of the effluent compounds, including dioxins and furans (Fig. 6-5). The wide range of furnace designs in use is undoubtedly responsible for the widely varying rates of dioxin emissions from the plants listed in Table 6-1. The Chicago plant, for example, emits only 42 billionths of a gram of dioxin per cubic meter of gaseous effluent from the stacks, while the Hampton (VA) plant emits over 4250 billionths of a gram per cubic meter of gaseous effluent. Clearly, these tremendous differences suggest the advisability of studying the design and operating conditions of such plants in an attempt to determine what design and operating parameters should be incorporated into future plants.

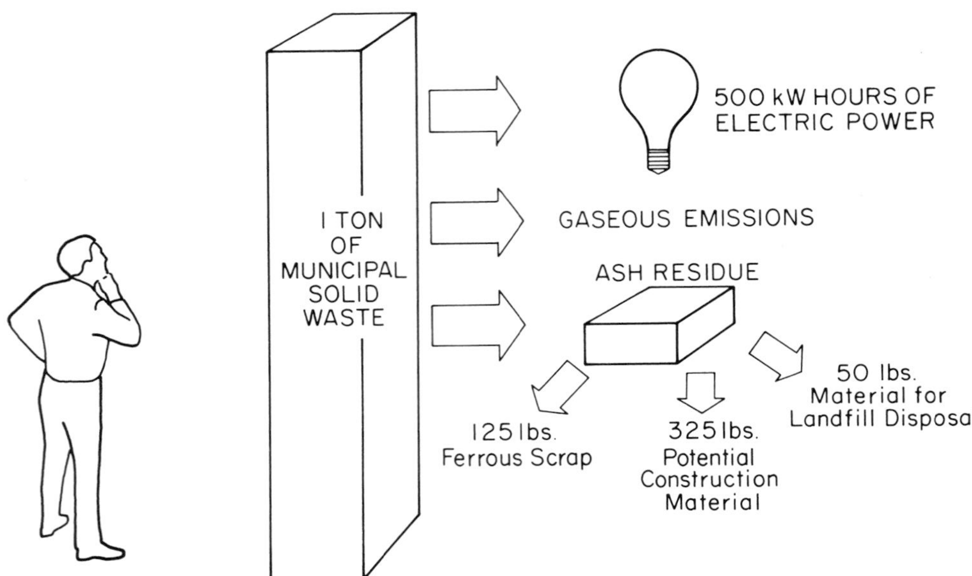


Figure 6-4. Results of incinerating one ton of municipal solid waste.

---

The three T's for good combustion are time, temperature, and turbulence.[3] The longer an element of solid waste is subjected to high temperature, the more complete is the combustion. The higher the temperature for a given exposure time, the better the combustion. For a given temperature and exposure time, the more turbulence the material is subjected to, the more complete the combustion. Good furnaces must therefore ensure a high temperature, a high degree of turbulence and a sufficiently long exposure time.

Given the varied composition of municipal solid waste and the need for resource recovery facilities to operate reliably throughout the year, it is important that there be an effective monitoring of furnace conditions. Three such methods are employed in modern facilities. One is the monitoring of emissions of carbon monoxide, certain

**Table 6-1** **Dioxin (PCDD) emission rates for a variety of operating resource recovery facilities.* Units are in nanograms of dioxin per cubic meter of gaseous effluent.**

| FACILITY (Country) | EMISSION RATE ($ng/m^3$) ALL PLANTS | HEAT RECOVERY PLANTS |
|---|---|---|
| STAPELFELD (Germany) | 31 | 31 |
| CHICAGO N.W. (USA) | 42 | 42 |
| ESKJO (Sweden) | 73 | 73 |
| STELLINGER MOOR (Germany) | 101 | 101 |
| PEI (Canada) | 107 | 107 |
| ZURICH (Switzerland) | 113 | 113 |
| BORSIGSTRASSE (Germany) | 128 | 128 |
| COMO (Italy) | 280 | 280 |
| ALBANY (USA) | 316 | 316 |
| DANISH RDF (Denmark) | 316 | 316 |
| ITALY 1 | 475 | |
| ITALY 6 | 569 | |
| BELGIUM | 680 | 680 |
| ITALY 5 | 1020 | |
| ZAANSTAD (Holland) | 1294 | |
| VALMADRERA (Italy) | 1568 | 1568 |
| HAMILTON (Canada) | 3680 | 3680 |
| HAMPTON (USA) | 4250 | 4250 |
| ITALY 4 | 4339 | |
| TORONTO (Canada) | 5086 | |
| ITALY 3 | 7491 | |
| ITALY 2 | 48,808 | |

*Source: Kay Jones, Roy F. Weston, Inc., Courtesy BFI, Inc. Plants are arranged in increasing order of emission of PCDD.

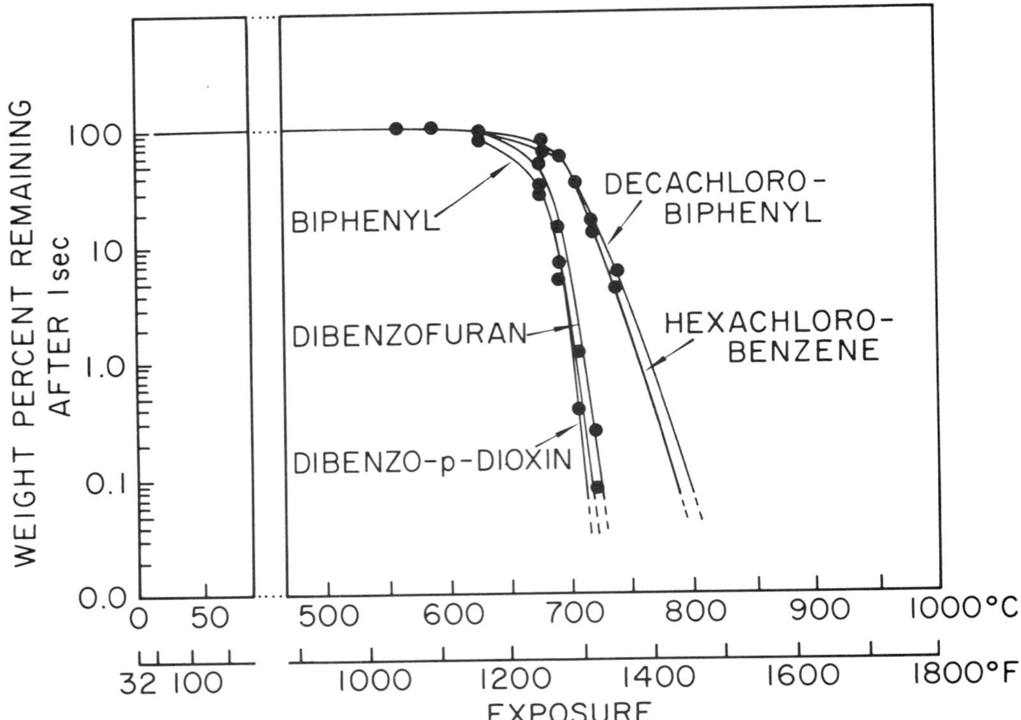


Figure 6-5 Destruction efficiencies of various compounds as a function of temperature. (Source: "Air Pollution Control at Resource Recovery Facilities", California Air Resources Board, May 24, 1984)

---

hydrocarbons and oxides of nitrogen. Another is the monitoring of the amount of carbon or combustible material remaining in the ash residue. A third is the monitoring of the efficiency of the boiler surrounding the furnace. The better the combustion, the lower is the emission of the foregoing gases, the lower the carbon and combustible material in the ash, and the greater is the boiler heat generated.

Of all of the surrogates for monitoring the performance of a furnace, carbon monoxide (CO) may be the most useful. The presence of high levels of CO in the flue gas indicates the presence of a significant amount of unburned carbon. Moreover, there is evidence that the amount of CO in the effluent is correlated positively with the amount of dioxins/furans emitted. Good furnace design and operation should keep CO levels in the effluent below 100 parts per million ($mg/m^3$).

The following general guidelines appear to foster good combustion[4]:

1. The grate (stoker) should be covered with fuel (by a uniform depth of garbage and trash) across its width. The depth at any given location on the grate should be consistent with the air deliverable for combustion at that

point.

2. There must be an air distribution system that will apportion air to the proper burning rate of waste along the entire breadth and width of the grate.

3. Underfire air should be introduced in a carefully controlled manner. Depending upon the particular technology it may be concentrated in a small area or spead over a large area. Zones of high pressure air and "blowtorch" effects should be eliminated. Bursts of air in one section of the fuel bed prevent the even mixing of air into the burning refuse in other areas.

4. Air must be introduced into the burning refuse both above and below the burning bed. Oxygen provided through the overfire system helps to complete the combustion of any hydrocarbons that were not oxidized near the fuel bed.

5. Steps must be taken to prevent the buildup of slag within the furnace. Slag can damage the boiler system, and also result in poor combustion by preventing proper air mixing into the fuel bed.

6. Gases generated in the incineration process should experience maximum mixing, to enhance the chances that oxygen will come into close proximity to any unburned particles, as well as to provide maximum dwell time of the gases before release to the atmosphere.

## 6-4 EUROPEAN INITIATIVES IN RESOURCE RECOVERY TECHNOLOGY

The methods chosen by a country to deal with societal problems are determined as much by the economic realities as by the state of technology. In Europe, where the costs of energy historically have been high relative to those in the United States and where land for dumping purposes has been scarce, there has been a much more aggressive adoption of resource recovery incineration methods for disposing of garbage and trash than in the United States. But the picture is changing in the United States as land becomes scarcer, environmental problems of landfills increase and resource recovery technology improves.

A critical element in the efficient operation of a resource recovery facility is the design of the stoker-grate system used in the furnace. It is not difficult to see how important the grate structure is since, clearly, it is not possible to form a massive pile of garbage, strike a match, and stand back and watch complete combustion take place. This is especially true because of the complex mix of materials in the typical solid waste stream, ranging from material with high moisture content, such as freshly cut grass, to carpets, wood and even large chunks of metal, including furniture and washing machines. The keys to hot and uniform combustion are constant mixing of air into the material being burned, and the use of partially combusted material to heat and ignite the new material introduced into the combustion chamber.

Three major European grate designs have found world-wide application. One design, the Martin System, has a reverse reciprocating grate system; another, the VKW System, has a series of rotating drums as a grate; and the third, the Von Roll System, has a reciprocating grate.

Figure 6-6 shows the Martin process. In this design the grate has a reverse reciprocating action: it moves alternately down and back to provide continuous motion of the refuse. The net motion of the refuse is downward toward the bottom of the furnace but the agitation caused by oscillation of the grate causes considerable mixing of burning refuse with newly introduced material, leading to rapid ignition and uniform burning.

In the VKW process (Fig. 6-7) large rotating drums slowly move the refuse toward the bottom of the furnace. This system also utilizes the ruffling of the garbage and the injection of air to enhance combustion.

Figure 6-8 illustrates the Von Roll system. There are three grate sections in this design: the first to dry the newly-introduced refuse and ignite it; the second to serve as the primary combustion grate; and the last as the stage on which the refuse is reduced to ash. Grate elements move in such a way that at a given time for any pair of elements, one is moving and one is stationary. Such a design results in the refuse moving slowly toward the bottom of the furnace but the shuffling action of the grates agitates the fuel bed enhancing significantly the combustion process.

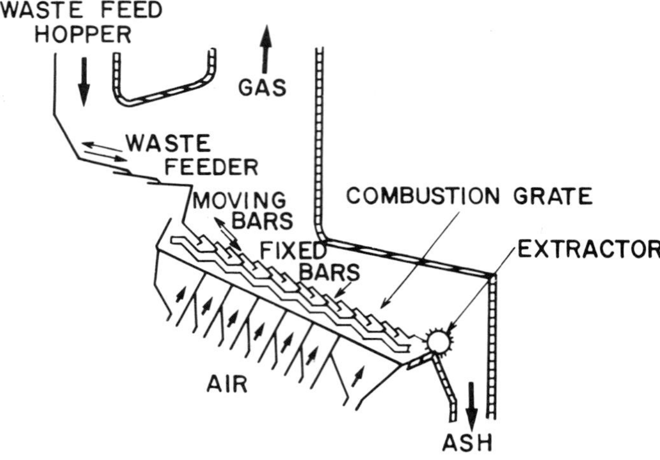


Figure 6-6 Grate system for Martin resource recovery incinerator.

Older American incinerator systems did not involve the agitation-generating features found in European systems. Instead, a series of two or three traveling grates was employed; drying took place on the first section and full or partial combustion on the second section and also on the third section, if present. Refuse entered the grates from a charging chute and was slowly carried through the various stages of drying and combustion with the residual ash discharged on a belt collection system. There now are a number of American systems of design similar to that of the European systems but, in general, they are not as sophisticated.

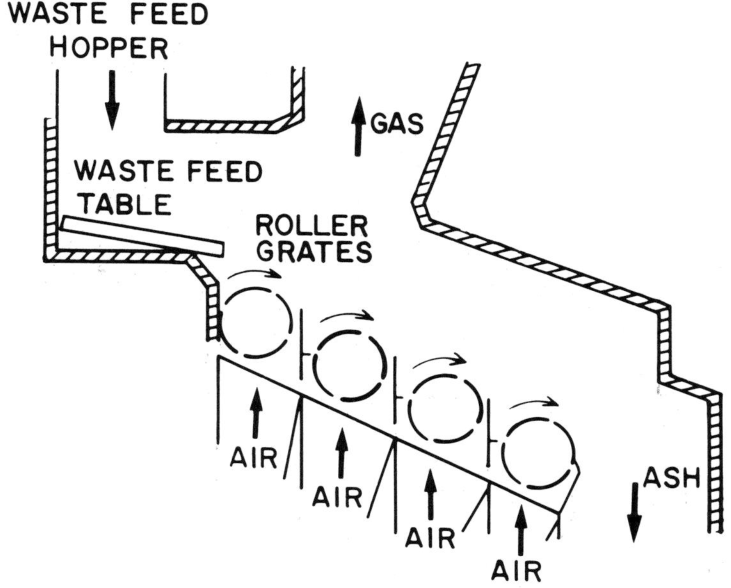


Figure 6-7 VKW grate system for resource recovery incinerator

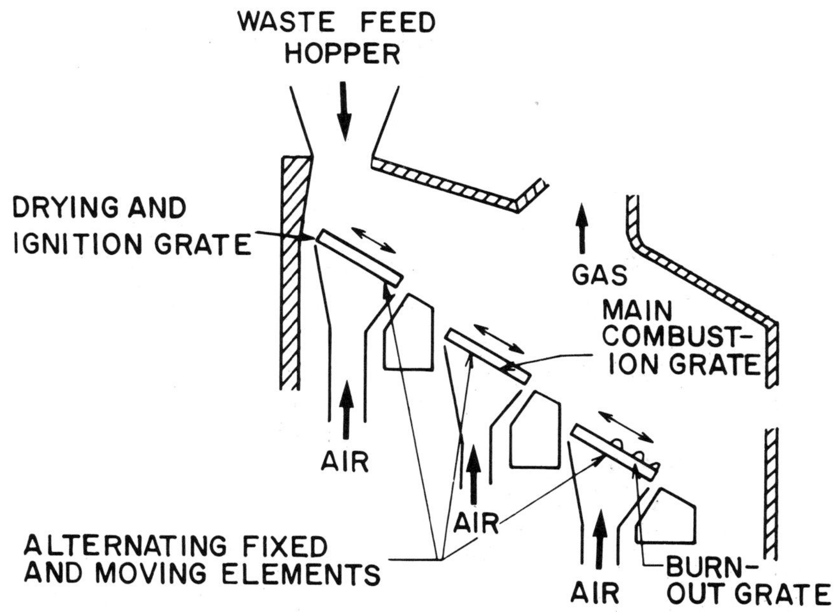


Figure 6-8 Von Roll grate system for resource recovery incinerator.

In another technique employed at some United States facilities, waste is pneumatically injected into the furnace system and burned while suspended in the furnace chamber, rather than being burned completely on a grate. To make this process efficient, the injected refuse must be largely free of noncombustible material and reduced to relatively fine elements, since large sections of any material introduced into the furnace will not burn before falling to the floor of the furnace. Thus, the refuse must be processed to be used in such facilities. The product of the processing step is known as refuse-derived fuel (RDF). In particular, some method of shredding material must be employed before the refuse is fed into the combustion chamber. Shredding can be problematic, since many items do not lend themselves to this process and, in addition, danger to the operator exists when shredding potentially explosive items. Processing normally includes magnetic extraction of bulk ferrous metals, and a screening step to remove fine glass particles and grit, which can cause slagging in the furnace. In contrast, European systems are designed to accept essentially all items without any processing. The shredding concept has not had a great deal of success in the United States and only a few plants remain in operation. Many have been closed because of serious mechanical and economic problems.

## 6-5 AMERICAN ENTRY INTO RESOURCE RECOVERY FIELD

It is interesting to review the origins of the growing utilization of mass incineration methods for disposing of solid wastes in the United States. In the 1970s a consensus developed that something had to be done to identify new methods of disposing of municipal wastes, to replace or, at least, to augment the use of landfills and small-scale volume-reduction incinerators. At the time, and in many cases now, landfills were little more than open dumps which were growing ever higher in elevation, leading in many cases to the poisoning of surface and groundwater supplies. Apartment-type incinerators then in use emitted large quantites of smoke and noxious odors, making life in heavily populated areas less and less desirable.

Against this background the then recently formed United States Environmental Protection Agency set about to determine which direction should be taken by the country in dealing with its solid waste disposal problem. To make sure that the technology identified represented the state-of-the-art, the agency funded a number of research/demonstration projects. There was very little attention paid to the systems already operating in Europe. It was assumed that since the European systems were designed many years before, they were essentially irrelevant in the selection of designs for the future decades. Millions of dollars were assigned by the EPA for a research and development effort that led to the development of systems such as those used in Hempstead, New York; Bridgeport, Connecticut; Hampton, Virginia; Franklin, Ohio; Baltimore, Maryland; San Diego, California; and St. Louis, Missouri. These systems were essentially failures, although the nature of their design flaws became known only after large sums in construction and operation had been spent. Because of the lack of ongoing, solid research in the years before, the pressures to identify quick solutions to the mounting garbage disposal problem led some municipalities to adopt technologies that were far from proven, and far inferior to the European systems which had evolved over the decades before.

Many of today's operating resource recovery plants which are experiencing various emission and combustion efficiency difficulties are based on designs developed during the past decade in the foregoing effort to find an improved incineration method. The lessons learned in this process underscore the stubborness of the waste disposal problem, and the need for long-range research, development and planning.

## 6-6 AIR POLLUTANTS IN INDUSTRIALIZED SOCIETIES

In any urban environment the atmosphere contains many pollutants. Some are relatively innocuous but unpleasant; others pose potential health hazards. In considering the impact of a routine, ongoing process such as mass incineration of solid waste, it is important to examine the extent to which the resulting emissions may add to the existing burden of atmospheric pollution, both in absolute and in relative terms.

In defining air quality one normally refers to the National Ambient Air Quality Standards (NAAQS) established by the federal government under the Clean Air Act. These standards, described in more detail in Appendix 2, utilize the levels of sulfur dioxides, nitrogen dioxides, lead, photochemical oxidants, carbon monoxide, and particulate matter as criteria in assessing air quality. These compounds are thus called the "criteria pollutants". There are other pollutants regulated by the Clean Air Act, but since they are not utilized as criteria for defining the quality of air, they are referred to as "non-criteria pollutants". They include asbestos, beryllium, mercury, fluorides, vinyl chloride, sulfuric acid mist, hydrogen sulfide, and reduced sulfur and sulfur compounds.

### Major Atmospheric Pollutants

One of the principal pollutants of the atmosphere is carbon monoxide, a gas produced whenever any carbon-containing material is burned. Carbon monoxide is more than 10 times as prevalent in the atmosphere as any other single pollutant. Of the 200 million tons of pollutants emitted into the atmosphere in the United States each year, almost half is carbon monoxide.[5] Approximately two-thirds of the total comes from automobile exhaust. In the city of Los Angeles alone, over 10,000 tons of carbon monoxide are generated and released into the atmosphere from automobiles every day.[6]

The inhalation of carbon monoxide in large concentrations can be life threatening. Carbon monoxide can pass easily into the lungs and then directly into the blood stream. Once there, it attaches to hemoglobin, and greatly reduces the efficacy of hemoglobin as a carrier of oxygen. The body senses a reduction in the oxygen being delivered by the blood; the heart rate is elevated to compensate for the loss; and the breathing rate is stimulated. If, in the process, more and more carbon monoxide is inhaled, life support mechanisms can degenerate quickly. Indeed, at carbon monoxide concentrations of 1,500 parts per million, human life is threatened. Any activities which add to the CO burden of major metropolitan areas need to be evaluated carefully.

The proposed Brooklyn Navy Yard 3,000 tons per day (TPD) resource recovery facility is expected to produce approximately 366 tons of carbon monoxide per year (See Table 6-2 and Ref. 7). This sum is to be compared with the New York City total carbon monoxide emissions of 644,208 tons per year. An industrial coal boiler producing the same amount of electricity as the proposed Brooklyn Navy Yard resource recovery facility would emit approximately 187 tons of carbon monoxide per year. The additions of CO from a 3,000 ton per day resource recovery facility would amount to about .03% of the total CO burden.

The next most abundant pollutants in most cities are the sulfur compounds. In industrial areas the concentration of sulfur dioxide is commonly in the range of 1.7 to 3.0 parts per million--close to the threshold level for human detection by smell. The effects of inhaling sulfur dioxide can range from discomfort as a result of the production of sulfuric acid in the throat and lungs (by the chemical reactions of the dioxides with moisture in the respiratory system), to more serious complications which arise when airborne particles with sulfur dioxide adsorbed onto their surfaces are inhaled and penetrate into lung tissue. Such deep penetration is thought to be capable of causing emphysema and bronchitis.

Table 6-2 **Comparison of projected emissions from the Brooklyn Navy Yard Incinerator, a steam generating facility and a coal boiler producing the same amount of electricity. Emissions stated in tons per year [Ref. 1].**

| Pollutant | Brooklyn Navy Yard Resource Recovery Facility | Hudson Ave. Steam Generating Facility | Industrial Coal Boiler |
|---|---|---|---|
| Particulate Matter | 161 | 128 | 486 |
| Sulfur Dioxide | 1177 | 1435 | 1847 |
| Nitrogen Dioxide | 2973 | 1300 | 3403 |
| Carbon Monoxide | 366 | 132 | 187 |
| Hydrocarbons | 66 | 26 | 56 |

The emission level of sulfur dioxide from the proposed Brooklyn Navy Yard resource recovery facility is projected to be 1,177 tons per year. For comparison, total production of sulfur dioxide for New York City is 56,336 tons per year. An industrial coal-fired boiler producing a comparable output of electricity would produce approximately 1,847 ton per year, or 670 tons per year more than the proposed resource recovery facility.[8] The levels of sulfur dioxide and hydochloric acid, which is also generated in large quantities in the incineration process, are significantly reduced through the use of dry scrubbers and baghouse filter- electrostatic precipitator systems, where an atomized chemical reagent spray serves to remove acid gas from the effluent gas stream.

Another class of criteria pollutants is the nitrogen oxides. Overall, these are produced at much greater rates by natural sources than by human activities. But in heavily populated areas, the anthropogenic production can dominate because combustion of fossil fuels is one of the primary mechanisms of production.

The typical level of nitrogen dioxide in an industrial area is 1 part per million. Exposures to levels of 50 ppm are quite hazardous, and methemoglobinemia, a change in blood chemistry, occurs at exposure levels of 100 ppm.[9]

Although sulfur dioxide and nitrogen dioxide cause illness in similiar ways, the absence of the warning smell which sulfur dioxide provides makes nitrogen dioxide especially dangerous. Individuals unaware of the impending danger may remain in areas of high concentrations of nitrogen dioxide longer than they otherwise would.

Nitrogen dioxides also have been implicated in the formation of smog and acid rain. In the atmosphere, $NO_2$ may be oxidized and combined with water vapor to form nitric acid ($HNO_3$), which contributes to acid rain.

The proposed Brooklyn Navy Yard facility is expected to produce 2,973 tons of nitrogen oxides per year. The total New York City production is 191,205 tons per year. An industrial coal-fired boiler which produces the same output of electricity as the proposed Brooklyn Navy Yard facility would release 3,403 tons or nitrogen oxides per year.[10]

Lead is used in a wide variety of products: plumbing pipes, radiators and batteries, paints, printing inks, glass, pottery, and electronic components, to mention just a few. The combustion of gasoline in automobiles is a primary cause of elevated lead concentrations in the atmosphere and this is the basis for the recent attention given to the regulated use of unleaded gasoline in newer model cars. Lead in the environment is a source of public health concern because it compromises the ability of the body to form hemoglobin and it can seriously damage the central nervous system, the kidneys and the reproductive system. Comparison with archaeological studies indicates that the lead level in bones of "modern industrial man " is more than 100 times higher than that of humans who lived 4,500 years ago.

The concentration of lead in the blood of the average human is about 10 to 15 micrograms per deciliter of blood. Lead poisoning can occur at about 50 micrograms per deciliter of blood. Urban atmosphere contains lead in concentrations normally much less than 10 micrograms per cubic meter of air and is not believed to pose any significant threat to adults even when exposed over long periods of time. However, individuals who work in certain jobs such as lead smelting and battery manufacturing are known to be exposed to levels ranging as high as 900 micrograms per cubic meter and have been known to develop lead blood levels of 90 micrograms per deciliter.[11]

The proposed Brooklyn Navy Yard resource recovery facility is expected to emit 14.5 tons of lead annually.[12] Data on lead emissions citywide are not available at this time. The principal contributors of lead to the urban atmosphere are automobiles and industrial processing. Emissions from a coal-fired power plant producing the same electricity output as the proposed Brooklyn Navy Yard facility and equipped with modern emission control technology would be approximately 3 to 5 tons of lead per year.[13]

## 6-7 SPECIFIC EMISSION ISSUES FOR RESOURCE RECOVERY FACILITIES

### Metals

Metals that are emitted from resource recovery plants in relatively high levels on fly ash particles include lead, cadmium, zinc, copper, manganese, silver, mercury, and tin.[14] Most metals are enriched on the smaller fly ash particles (less than 2 microns in diameter). This observation can be explained by the volatilization of metals during the combustion of refuse, and subsequent condensation at lower temperatures and adsorption onto the finer sized particles, which have greater surface area per unit volume available for reaction than do larger particles. Furthermore, the presence of a higher concentration of fine particles in the flue gas will increase the probability that a volatilized metal will condense onto particles. Of the heavy metals, mercury is the only one that does not show a high degree of affinity for adsorption onto fine particles. This behavior is the result of the high vapor pressure of mercury. Approximately half of the mercury in the flue gas is in the vapor phase at temperatures characteristic of flue gas.

Identification of the sources of some of the more toxic metals in the refuse, and removal prior to combustion may decrease the emissions of some of these metals.[15,16] Trace elements found in urban refuse are listed in Table 6-3. Printing inks have been found to be sources of lead, cadmium and zinc. Other metals used extensively in publishing are titanium, molybdate, magnesium, iron, and barium. Paints contribute lead, titanium, and chromium to emissions.

Cadmium and copper are concentrated in heavy combustibles like heavy-gauged plastics. Also, plastic stabilizers are sources of tin, lead and cadmium. Those metals in emissions that appear to be contributed in roughly equal amounts by the combustible and noncombustible fractions of the refuse include cadmium, chromium, lead, manganese, silver, tin, and zinc. These metals are associated with coatings of galvanizing materials, solders, pigments, and other surface agents, or with thin foils or wires in the noncombustible fraction. Metal emissions believed to be derived largely from the combustible fraction are copper, cadmium, mercury and magnesium. Lead in emissions is believed to be significantly derived from noncombustible sources (e.g. bulk metals). Therefore, removal of the noncombustible sources before incineration could effect some reduction in the emission levels of cadmium, chromium, lead, manganese, silver, tin, and zinc.

Following volatilization, major reductions of metal emissions can be accomplished by efficient collection of fly ash, particularly the finer size fraction, with electrostatic precipitators or baghouse filters. In general, reducing the flue gas temperature promotes condensation of many volatilized metals onto fly ash particles, which then can be removed by pollution-control devices--electrostatic precipitators and baghouse filters. Most systems are designed to reduce temperatures to below 500° F; temperatures below 250° F are even more effective.

The primary atmospheric pollutants described above are of direct concern to residents of urbanized regions and usually are monitored by the appropriate agencies. It has been demonstrated repeatedly that well-designed resource recovery facilities can meet the extant regulations governing the criteria polutants and, as discussed below, that the technology exists for satisfactorily controlling metal emissions. There are other important concerns, however, associated with the possible release of minute quantitites of highly toxic chemicals such as furans and dioxins as a result of the incineration of municipal solid wastes and other processes. At present there are no United States regulations governing such emissions, and the official position of the United States Environmental Protection Agency is that emissions of such hydrocarbons from properly operating resource recover facilities do not pose a health hazard. However, public concern remains. Some of the elements of the concern are discussed below.

**Table 6-3. Trace elements found in urban refuse**

| Element | Concentration (ppm dry weight) |
|---|---|
| Silver (Ag) | 7 |
| Aluminum (Al) | 5,978 |
| Barium (Ba) | 130 |
| Calcium (Ca) | 6,848 |
| Cadmium (Cd) | 9 |
| Cobalt (Co) | 7 |
| Chromium (Cr) | 65 |
| Copper (Cu) | 250 |
| Iron (Fe) | 1,630 |
| Mercury (Hg) | 1 |
| Potassium (k) | 913 |
| Lithium (Li) | 2 |
| Magnesium (Mg) | 1,087 |
| Manganese (Mn) | 250 |
| Sodium (Na) | 3,152 |
| Nickel (Ni) | 54 |
| Lead (Pb) | 674 |
| Antimony (Sb) | 22 |
| Tin (Sn) | 98 |
| Zinc (Zn) | 1,087 |

Source: Lowes, S., B. Hayne and W.J. Campbell. 1978. Pre-burn separation should limit metal emission. Waste Age 9:51-59.

The incineration and combustion processes and the byproducts of combustion of a diverse waste stream consisting of paper, wood, metals, glass and plastics are chemically very complicated. Though the mechanics of these processes are known, the science is not fully understood. Certain gases can be created directly by the chemical breakup of items during incineration. Other compounds may be created through the interaction of various compounds present in the burning environment. We review below issues regarding the production of some compounds of special interest -- PCBs, dioxins and furans.

## PCBs

Special concern has been expressed about the incineration of materials containing polychlorinated biphenyls (PCBs) because low concentrations of these chemicals have been shown to cause cancer in laboratory animals. This concern has extended to the incineration of municipal wastes which are known to contain numerous items fabricated with PCBs.

PCBs present an interesting example of the way in which products with potential adverse effects can become almost ubiquitious in their use by society. Through research with hydrocarbons, scientists discovered that chains of molecules of hydrocarbon coupled with chlorine produced material that had unique electrical insulating properties and good stability as a plastic. Almost immediately these chemicals, PCBs, found their way into adhesives, fireproofing materials, electronic components, and paints, as well as carbonless paper. The carbonless paper contained PCBs in Aroclor 1242 microcapsules. Although the discovery of the potential health hazards of PCBs led to suspension of manufacture in 1971, they by then were distributed widely in the environment and will be with us for a very long time; at least for decades. As an example, the rapid market success of PCB-laden carbonless paper, and the recycling of this paper after use, has led to PCBs being present in essentially all paper products using recycled paper. Thus, there are clear reasons for being concerned about what effects landfilling and incineration will have on the further dispersal of PCBs in the environment.

Studies to determine what occurs when PCB-containing materials are burned, and studies to determine the environmental conditions which promote the formation and release of PCBs are limited in scope. One scientific paper reported that the level of PCBs in the effluent of a resource recovery facility was essentially unrelated to the PCB content of the materials incinerated and that it was consistent with the *de novo* formation within the plant[17] Other observations have indicated that PCB levels in residential dwellings were higher than those outside because of outgassing of caulking compounds, small electrical equipment, and ballasts of faulty fluorescent lights. In any case, existing data do indicate that at least 99.9 percent of the PCBs contained in municipal solid waste are destroyed by the high temperature incineration process characteristic of modern resource recovery facilities.[18]

## Dioxins and Furans

A chemical compound having two benzene rings linked by two oxygen bridges in the presence of specific chlorine atom arrangements is called a polychlorinated dioxin. Another related group of compounds are the dibenzofurans (frequently called simply furans). The principal difference between furans and dioxins is the presence of one oxygen bridge in furans, rather than the two in dioxins. The term "bridge" refers to the the chemical bonding provided by oxygen; since an atom of oxygen is divalent (i.e., it is able to form two electron bonds), it can bind two benzene molecules.

To appreciate the features that characterize the dioxin and furan families, it is useful to examine diagrams of their respective chemical structures. First, it is necessary to introduce the shorthand notation used by chemists in describing such compounds. The furan and dioxin families of chemicals contain the elements carbon (C), hydrogen (H), oxygen (O), and chlorine (Cl). Carbon atoms, which can bond to each other to form complex skeletons, in this case are bonded into 6-membered rings that are fused together to form molecules. In diagrams using the chemical shorthand notation the carbon atoms (C) are usually not explicitly written out; the bonds alone are drawn to indicate the carbon skeleton. The hydrogen atoms (H) are also omitted, as are their bonds. A 6-carbon ring with a circle inside is called an aromatic (benzene) ring, and it denotes particular characteristics to a chemist. Thus, the abbreviated notation used in Fig. 9 (a) is equivalent to the dioxin diagram shown in Fig. 6-9 (b), which illustrates not only the carbon ring but the hydrogen links as well.

Using the abbreviated notation the difference between dioxins and furans can be readily displayed, as in Fig. 6-10. In the case of dioxins, two oxygen atoms serve to bind the benzene rings. In the the case of furans the binding takes place through one oxygen bond and one direct bond. One should also note from this figure that each of the carbon sites not participating in the binding bears a specific labeling number. This labeling scheme permits a unique distinction to be made between different members of the dioxin or furan families. For example, one dioxin of great interest is the so called 2,3,7,8 TCDD. The "TCDD" stands for tetrachloro-dibenzo-p-dioxin, or just "four chlorine atoms-two-benzene ring" dioxin of type p. The "2,3,7,8" refers to the fact that this dioxin has chlorine atoms attached at sites 2,3,7 and 8. Dioxins with chlorine atoms bound to other sites will, of course, have different designations. The family of dioxins is shown in Fig. 6-11, with the 2,3,7,8 TCDD member being the third from the top. Examples of members of the furan family are shown in Fig. 6-12.

Dioxins appear to retain their chemical integrity up to temperatures of 1,300° F; above that they disintegrate through the breaking of the various bonds (see Fig. 6-5). At standard atmospheric temperature and pressure, the solubility of dioxins in water is quite low, and their vapor pressure is also low, indicating that very little of the material volatizes at ambient temperature and pressure.

The Tetrachloro-dibenzo-p-dioxin (2,3,7,8 TCDD) is the most toxic dioxin discovered so far. Although there has been no known case of long-term human disability, or death, resulting directly from exposure to dioxin, there is a body of research which indicates that this compound is very toxic to some small animals. Indeed, it currently is the most toxic of all synthetic chemicals tested on animals. In terms of lethal effects for each gram of toxin per kilogram of body weight, TCDD dioxin is 10,000 times as potent as cyanide. In comparison with the dreaded Botulism Toxin A, however, it is less than one-ten-thousandth as potent.[19] Table 6-4 illustrates the effect of dioxin on various animal species. No broadly accepted explanation exists for why dioxins are so much more lethal to certain species than to others. These discrepancies further demonstrate just how difficult it is to predict the effects on humans.

Carcinogenicity of TCDD, as well as effects on the reproductive systems and the fetuses, has been demonstrated in small laboratory animals. There are no experimental data on the effects of dioxins on humans. Since dioxins are not naturally occuring, at least not in significant quantities, the epidemiology of effects over time is not available. Industrial accidents have been infrequent and have not led to diagnostic insights of the precise effects of exposure. The incidents that have been examined indicate that exposure is followed by the skin disorder, chloracne, and short-term liver damage.[20] Long-term studies and detailed analyses will be required to assess the true carcinogenic effects on humans.

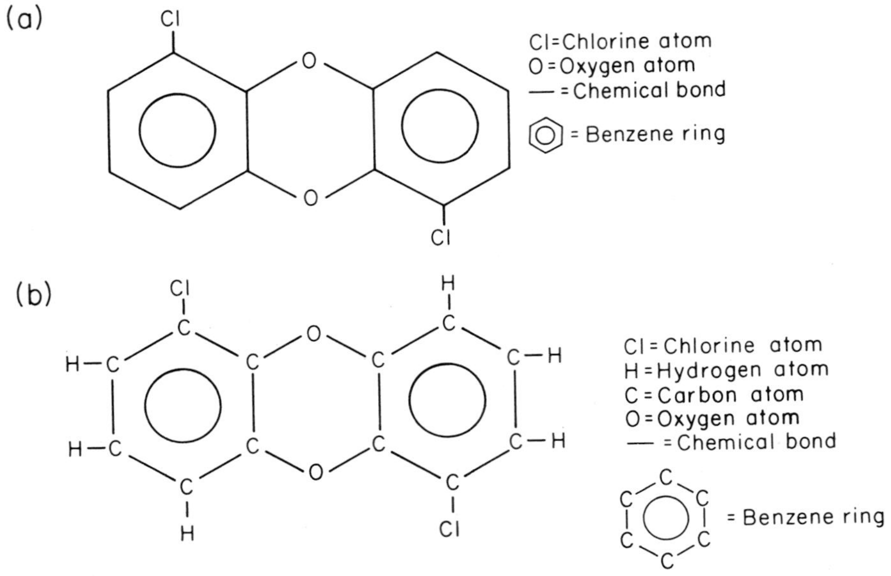

(a)

Cl=Chlorine atom
O=Oxygen atom
— =Chemical bond
= Benzene ring

(b)

Cl=Chlorine atom
H=Hydrogen atom
C=Carbon atom
O=Oxygen atom
— =Chemical bond
= Benzene ring

Figure 6-9: (a) Chemical diagram of dichloro-dibenzo-p-dioxin using shorthand notation. (b) Diagram of same chemical with carbon ring and hydrogen bonds illustrated.

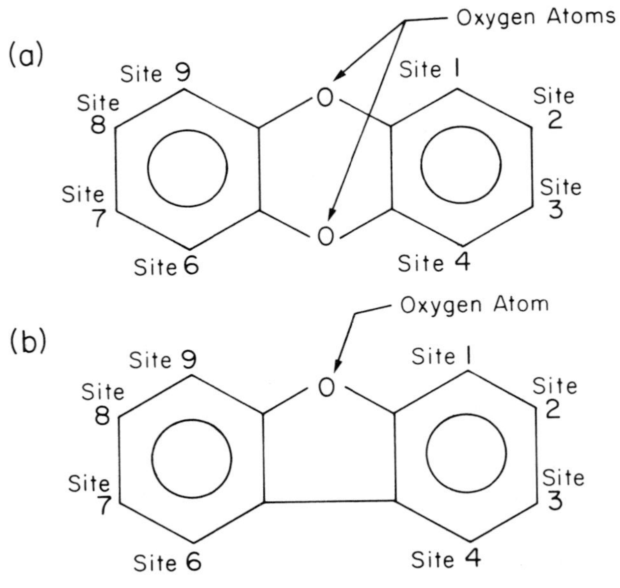


Figure 6-10 (a) Generic diagram of dioxins. (b) Generic diagram of furans.

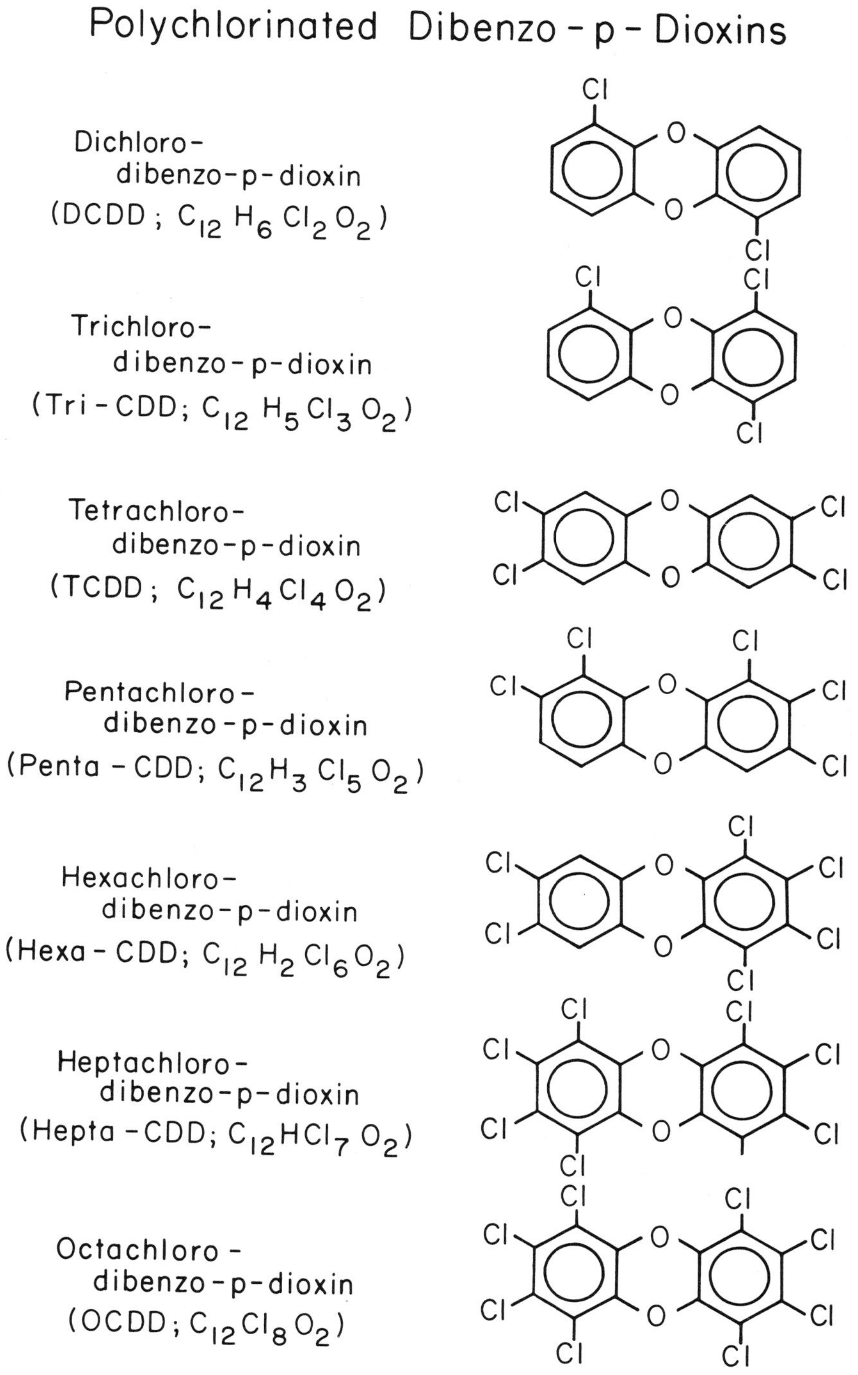

## Polychlorinated Dibenzo-p-Dioxins

Dichloro-dibenzo-p-dioxin (DCDD; $C_{12}H_6Cl_2O_2$)

Trichloro-dibenzo-p-dioxin (Tri-CDD; $C_{12}H_5Cl_3O_2$)

Tetrachloro-dibenzo-p-dioxin (TCDD; $C_{12}H_4Cl_4O_2$)

Pentachloro-dibenzo-p-dioxin (Penta-CDD; $C_{12}H_3Cl_5O_2$)

Hexachloro-dibenzo-p-dioxin (Hexa-CDD; $C_{12}H_2Cl_6O_2$)

Heptachloro-dibenzo-p-dioxin (Hepta-CDD; $C_{12}HCl_7O_2$)

Octachloro-dibenzo-p-dioxin (OCDD; $C_{12}Cl_8O_2$)

Figure 6-11 Members of the dioxin family.

Tetrachloro dibenzo-furan

Pentachloro-dibenzo-furan

Hexacloro-dibenzo-furan

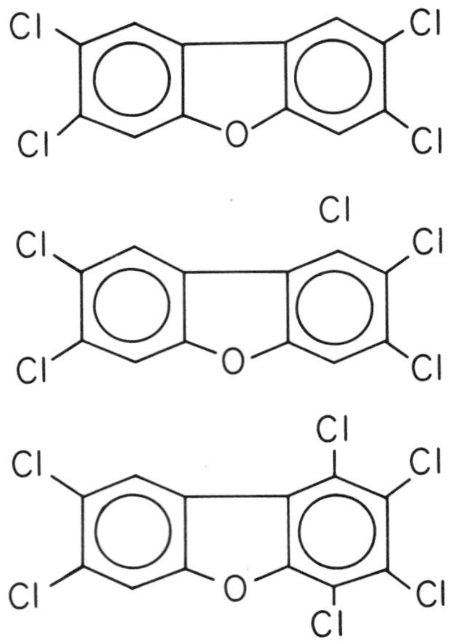

Figure 6-12. Examples of members of the furan family.

Dioxins are released in the environment by residential fireplaces and grills, incinerators, diesel truck mufflers, and the soil. Table 6-5 shows approximate levels of dioxin in parts per billion in various sites.[21]

Parameters that influence the rate of dioxin emission from resource recovery facilities include combustion temperature, air mixture, exposure time and post-furnace conditions. Observations of municipal solid waste incinerators indicate that increasing the temperature of combustion generally decreases the total amounts of dioxins and furans produced. These observations at operating plants are consistent with the belief that greater thermal and oxidative destruction of PCDDs, PCDFs, and precursor compounds occurs at higher temperatures. An extensive analysis of dioxin emission rates for a number of combustion characteristics revealed a strong positive correlation between increases in PCDD/PCDF emission rates and decreases in minimal combustion temperature, and a weaker correlation with average furnace temperatures.[22] When temperatures fall below 932 °F, emission rates appear to be greatly enhanced, perhaps indicating a change in the production process such as the generation rate of organic precursor compounds. At temperatures at and above 1,800 °F (1,000 °C), dioxins and furans are still detected, but levels are significantly reduced.

Many laboratory studies have been used to examine the production of dioxins and furans, and their precursors, from combusting and pyrolyzing of both chemically related and unrelated substances. While there is rather general agreement on a correlation between PCDD/PCDF production and temperature, some investigators believe it is very difficult to relate laboratory reaction analyses to dioxin and furan generation from the incineration of heterogeneous refuse. More useful information may be obtained by analyzing particular chemicals (i.e. chlorine and chlorinated aromatic compounds) in municipal solid wastes to determine how they influence the

**Table 6-4 Amount of PCDD dioxin proving lethal to 50% of a population of animals, by animal species [Ref. 14]**

| Animal | Micrograms of dioxin per kilogram of animal body weight |
|---|---|
| Guinea pig | 1 |
| Rat (male) | 22 |
| Rat (female) | 45 |
| Monkey | <70 |
| Rabbit | 115 |
| Mouse | 114 |
| Dog | >300 |
| Bullfrog | >500 |
| Hamster | 5,000 |

production of dioxins and furans. The apparent lack of correlation between the rate of production of dioxins and the presence of chlorine in the waste stream seems to preclude a simple solution of the problem by reducing the chlorine content of the input wastes. More recent data indicate that post-combustion conditions may play a major role in controlling the emissions of dioxins and furans. As the emerging flue gas cools these compounds are thought to form on particle surfaces (e.g., fly ash). This possibility further emphasizes the value of utilizing high efficiency baghouse filters and electrostatic precipitators.

In reviewing the possible impacts of dioxin and furan emissions from resource recovery plants, it is important to take into account the fact that these compounds have finite lifetimes, and decay into much less harmful forms within periods of days to, at most, several months. This fact has been overlooked in many treatments of this topic. Some scientists argue that the toxic loading of the atmosphere from the operation of a high temperature incineration facility is well below the estimated hazardous level.[23] One key premise of this analysis is that the photochemical decomposition of 2,3,7,8 - TCDD has a half-life (the time required for the degradation of half of the molecules in a sample) of hours in the air and on leaves and grasses, and less than a year in soil. To estimate the overall environmental burden, one must then estimate the amount of TCDD produced during a time period comparable to the pertinent half-life and not integrate over all times. For example, if the appropriate half-life is one year in soil, to estimate the accumulation of TCDD in the soil from a nearby resource recovery facility, one should only take into account the amount of TCDD settling onto the soil from the plant's fallout over the period of a year, and not for 5 or 10 years. Most of the deposition from previous years will have decomposed into presumably less harmful components, as a result of chemical and biological processes.

Table 6-5 Levels of dioxins in soils at various sites.

| Sample Source | TCDD | HCDD | $H_7CDD$ | OCDD |
|---|---|---|---|---|
| Soil | | | | |
| Rural | * | * | * - .05 | * - .2 |
| Urban (Lansing, Mich.) | * | .03-1.2 | .03-2 | .05-2 |
| Urban (Chicago) | .005-.03 | .03-.3 | .1-3 | .4-22 |
| Dow Chemical (Michigan) | 1-120 | 7-280 | 70-3,200 | 490-20,000 |
| Dust | | | | |
| Dow Chemical Laboratory | 1-4 | 9-35 | 140-1,200 | 650-7,500 |
| Midland, MI | .03-.04 | .2-.4 | 2-4 | 20-30 |
| Detroit, MI | * -.03 | * -.3 | .3 - 4 | .1 - 4 |
| St. Louis, MO | .3 | 2 | 34 | 210 |
| Chicago, IL | .04 | * -.3 | .6-3 | 3-8 |
| Wastewater Treatment Sludge | | | | |
| Milorganite (Milwaukee) | .31 | 2 | 30 | 180 |
| Incinerators | | | | |
| Dow Powerhouse | 38 | 2 | 4 | 24 |
| Dow Rotary Incin. Stack | * | 1-5 | 4-100 | 9-950 |
| Dow Tar Burner | * | 1-20 | 27-160 | 190-440 |
| Nashville Incinerator | 7.7 | 14 | 28 | 30 |
| European Incinerators | 2-20 | 30-200 | 60-130 | 40-120 |
| Mufflers | | | | |
| Diesel Truck Muffler | .023 | .02 | .100 | .26 |
| Auto Muffler | * -.008 | * | .003-.01 | .02-.07 |
| Other sources | | | | |
| Home Fireplace Soot | * -.4 | .2-3 | .7-16 | .9-25 |
| Home electrostatic precip. | * | .004-.008 | .009 | .02-.05 |
| Charcoal Broiled Steak | * | * | * | .03 |

* Not detected

It is of interest to examine the amounts of dioxins and furans produced in the incineration process. For each million tons of municipal solid waste burned, approximately 63,000 tons of fly ash is produced. Electrostatic precipitators can successfully collect between 95 and 99 percent of the fly ash, with the remaining 1 to 5 percent escaping into the atmosphere with the flue gas. With high efficiency fabric filters, more than 99 percent of the fly ash can be trapped and removed. This fly ash would contain between 100 to 3,000 parts per billion of adsorbed dioxins.[24] Operating at 3,000 tons per day of refuse incineration, it is projected that the Brooklyn Navy Yard facility will emit approximately one twentieth of a pound of dioxins per year. These emissions are estimated to lead to a certain ground level concentration that depends on distance from the facility but these levels are well below those currently used by various agencies in establishing what is considered to be a safe exposure level.

For example, the Netherlands government has adopted as an acceptable level of intake of TCDD 1 nanogram per kilogram of body weight per day.[25] This translates into an acceptable ambient concentration of roughly 350 x $10^{-5}$ micrograms per cubic meter. Using a safety factor of 250 in arriving at a standard, the Netherland standard is then approximately 1.4 x $10^{-5}$ micrograms per cubic meter. Calculations by Fred C. Hart Associates indicate that the impact of the Brooklyn Navy Yard facility would be 2.13 x $10^{-9}$ micrograms per cubic meter, a number 6,600 times smaller than the Netherlands standard, which itself is 250 times smaller than the assumed acceptable level.[26]

## 6-8 GENERAL ASSESSMENT OF EMISSIONS FROM INCINERATORS

In December 1984 a group of scientists met under the auspices of the New York Academy of Sciences to discuss the emissions from resource recovery facilities.[27] General agreement was expressed that the emissions of sulfur oxides, metals, chlorides and particulates, could be controlled if appropriate existing technologies are employed. Sulfur dioxide emissions from the incineration of municipal wastes are relatively insignificant when compared to those from coal-fired plants. Mercury and other volatile metals vaporize at the temperatures of modern incinerators and end up in the gaseous discharge stream. As described earlier, lowering the flue gas temperature to below 500$^{o}$F promotes condensation of metals onto fine particulates which can then be removed. The chlorine present in the waste stream is converted in large measure to hydrogen chloride.

This general agreement among the participants at the meeting concerning the ability to monitor and control the emissions of sulfur oxides, metals, chlorides and particulates stands in sharp contrast to the controversy that existed on the amounts and effects of the more complex organic compounds--in particular, the polychlorinated dibenzo-p-dioxins and polychlorinated dibenzo-furans. One or more of the following processes were thought to be involved in the production of dioxins: the release of dioxins present in the input waste, the synthesis of dioxins from direct precursors (i.e., the modification of structurally similar compounds), or the *de novo* synthesis of dioxins from basic organic materials (i.e., extensive rearrangement of much simpler organic compounds). The first mechanism does not seem to provide the sole source of dioxins since data indicate that the amount of dioxin emitted exceeds the dioxin in the input stream. Although there is evidence to support formation through precursors and *de novo* synthesis, there is no conclusive evidence regarding the relative contribution of each of these processes.

In almost any attempt to determine the effects of trace amounts of toxic chemicals on humans there is the problem of knowing how to scale what is learned from intense doses of chemicals delivered to test animals for a short period of time to the more realistic problem of the impact of smaller doses to humans over a prolonged period of time. Finally, as noted above, there also is the question of whether results from animal studies can be applied directly in assessing impacts on human health. There is the possibility that, on a per volume or weight basis, humans may be either less sensitive or more sensitive to the effects of chemicals such as dioxins and furans.

The panelists involved in the New York Academy of Sciences conference pointed out that all analyses to date have confined their attention to the worst-case situation for cancer generation. This approach has clear defects.

In summary, the above conference, and a similar series of symposia held at the State University of New York at Stony Brook, have reached the conclusion that of the many issues of potential concern associated with the operation of modern resource recovery facilities, most seem to be solvable with existing technology. The remaining uncertainties with the production of dioxins and furans may soon be resolved by current studies. Proceedings of the January, 1986 Stony Brook symposium are included in Appendix 3.

## 6-9 VEHICLE EMISSIONS IN RESOURCE RECOVERY FACILITIES

Vehicular traffic around a resource recovery facility consists primarily of incoming loaded trucks, outgoing empty trucks, and a small volume of traffic associated with transporting recovered materials to dealers, and ash residue to nearby landfills.

The internal traffic at a resource facility requires very little fuel per ton of garbage compared with the fuel required per ton of garbage for the operation of a typical sanitary landfill. The difference, of course, is the extensive vehicle operation required for the distribution and covering of garbage and trash within a landfill.

## 6-10 ODORS AND VECTORS AND SURFACE BLOWN LITTER

Resource recovery facilities cause essentially no detectable increase in the odor level in the region surrounding the facility. Because the interior of the plant is kept at a pressure lower than that of the outside environment, no air from the garbage pit escapes to the outside. Moreover, since the outside air that enters the building and the fumes from the garbage tipping floor are drawn into the furnace to facilitate combustion they emerge as part of the flue gas which is odorless.

Since the garbage and trash is unloaded from the transporting trucks inside a closed building there is no problem of surface blown litter at a properly operated resource recovery facility.

## 6-11 WATER POLLUTION AT RESOURCE RECOVERY FACILITIES

The major use of water in resource recovery plants is to condition the ash residue to minimize the problem of dust. In the Martin process, for example, just enough water is used to condition the ash and almost all of it evaporates in the ash discharger assembly. Moreover, many plants are designed so that cooling, wash-down and boiler blow-down water is directed to the ash discharger. There is little or no waste water effluent from modern resource recovery facilities, and therefore there is no significant environmental impact on either surface or groundwater supplies from waste water.

## 6-12 OFFSET OF EMISSIONS FROM ALTERNATE ENERGY SOURCES

In most modern resource recovery facilities, the furnace walls are lined with pipes which carry water. In normal plant operation, heat from the burning refuse converts the water to steam at a temperature of approximately 750° F and a typical pressure of 600 psi. Under such conditions the steam can readily be used to drive turbines to create electricity, in the same way as in large coal or oil-fired electric generation plants in operation around the world. This component of the operation of modern resource recovery facilities draws upon proven technology, although minor innovations have been made in the placement of the water pipes and in other aspects of the operation. A production of between 400 and 500 kilowatt hours of electricity can be assured from the incineration of a ton of typical municipal waste. As shown in Fig. 6-13, the incineration of 1 ton of solid waste produces roughly the same electrical output as the burning of 1 barrel of oil. This is enough energy to operate 5,000 100-watt bulbs for 1 hour, to operate 500 hair dryers for 1 hour, or to provide energy to a typical apartment for one month. A 1,200 ton per day disposal facility can operate a 29 megawatt turbine, which can produce enough electricity to power more than 20,000 typical homes.[28]

In assessing the environmental impacts of resource recovery facilities, it is appropriate to note the positive contribution made by such facilities by reducing the amount of electricity that would have to be generated by some other means. In particular, consider the case of a 1,200 ton per day resource recovery facility that generates 29 megawatts of electricity. If 29 megawatts of electricity were generated by an oil-fired power plant, there would be emissions of specific quantities of various chemicals. In assessing the differential impact of a resource recovery plant on air quality, its total emissions should be reduced by an amount equivalent to what would be produced by a conventional electric generating station with the same electrical output, since these emissions can be viewed as having been avoided through the operation of the resource recovery plant. This presumes that the electricity generated by the resource recovery facility replaces an equivalent amount of electricity which would have had to be produced through other means.

## 6-13 ASH RECYCLING

Since the mid-1870s incineration of refuse has been in routine use for garbage and trash disposal in Europe. Ash from the incineration process has been utilized both as fill material and as aggregate in the construction of roads. With the increasing number of resource recovery facilities in the United States and the increasing complexity of the material being incinerated, the question of how the residual ash produced by the incineration process should be disposed of is emerging as a critical issue.

Approximately 20 to 25 percent of the original solid waste mass remains as ash after incineration. The ash has the approximate chemical analysis shown in Fig. 6-14. Although this reduction in mass is extremely valuable, there is still a need to provide an ultimate disposal for a significant fraction of the waste stream, and this need is currently met primarily through the use of landfills. Furthermore, to the extent that ash contains trace amounts of PCBs, PVCs, metals, dioxins and furans, and other compounds that are known to be potentially toxic, care must be taken in choosing how and where to dispose of the ash.

There are several promising uses of residue from incineration plants. Some examples include the construction of embankments, landfill cover, graded material in road pavements, and aggregate for cement and masonry manufacture for construction purposes, and for artificial fishing reefs.

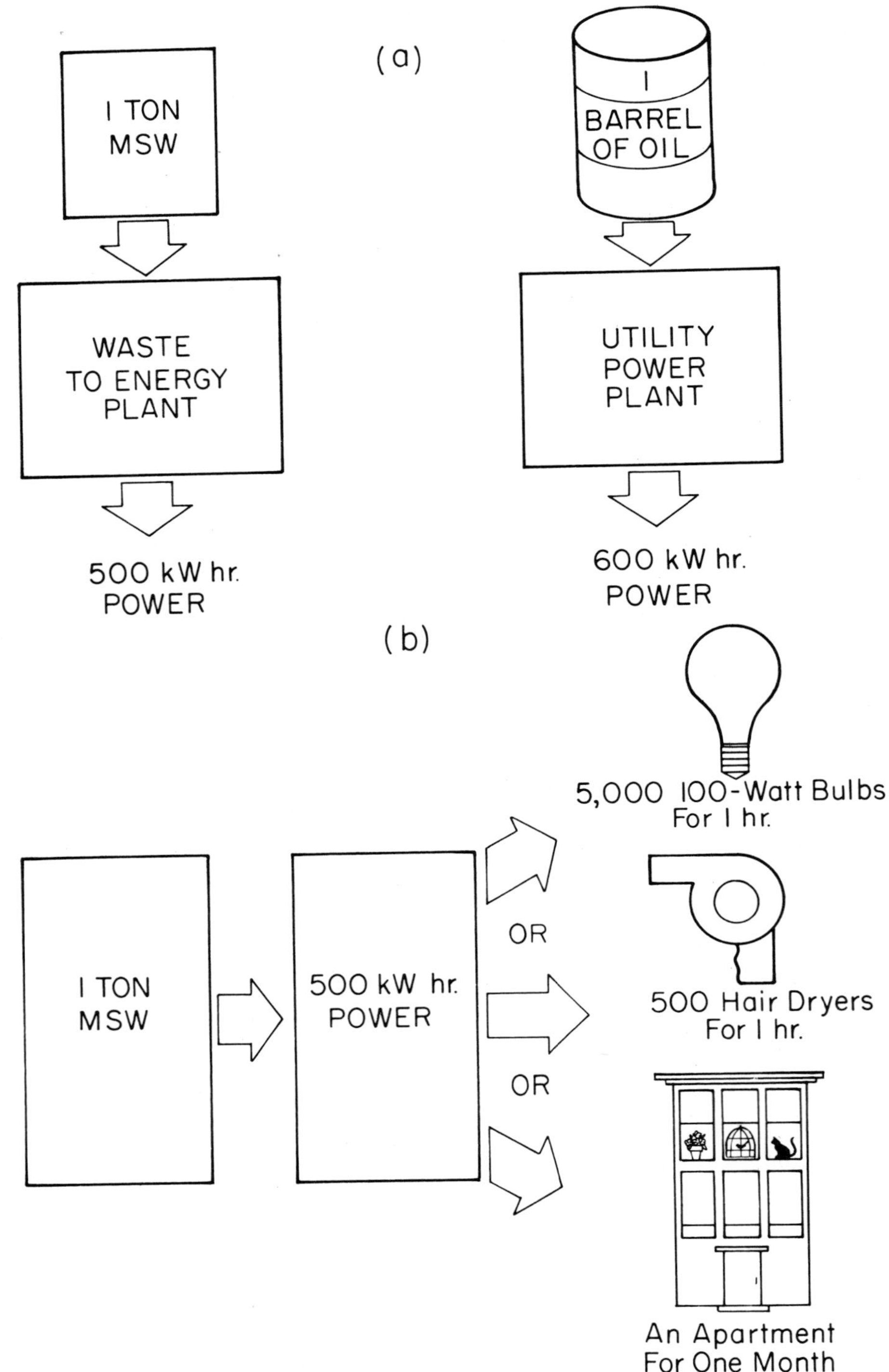


Figure 6-13 Electrical output from the incineration of one ton of municipal solid waste.

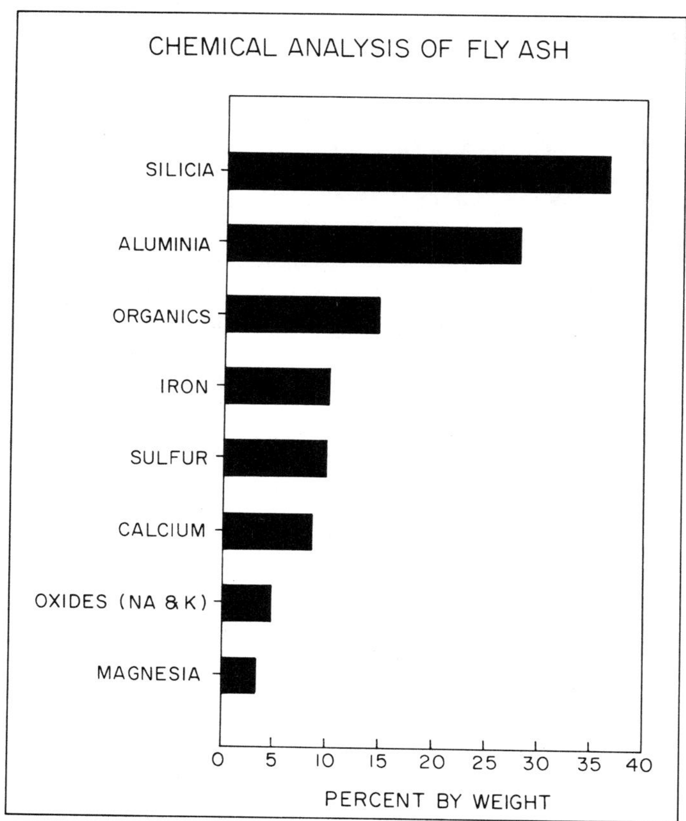


Figure 6-14 Chemical composition of fly ash.

---

In considering the use of ash residue for fabricating construction material, the structural rigidity and durability must be examined. Preliminary studies have pointed out certain difficulties. Differential expansion of the residual aluminum and glass products in the ash relative to the alkaline materials in the concrete blocks reduces the structural strength below acceptable levels for use as concrete cement. However, other research has demonstrated that ash from coal-burning plants can be used to make concrete blocks of adequate strength for use in making artificial reefs.[29] This work is being extended to ash from municipal incineration plants to determine what chemical stabilization may be required to bring concrete blocks made using fly ash to construction grade strength (see additional discussion in Chapter 5).

Ash from resource recovery plants contains varying quantities of ferrous and aluminum components. The ferrous metals can be separated from the ash stream by the use of magnets. Recovery of aluminum and ferrous scrap from the ash stream is not very profitable and, indeed, many of the original programs to recover these materials have been suspended for economic reasons. Problems encountered include the variable quality of the recovered products, the need for extensive additional processing to ensure quality and meet product specifications, and the lack of markets for the recovered materials.

Figure 6-15 illustrates an example of an operating system that currently segregates aluminum and ferrous material from the ash stream. Residue is first moved from the ash dischargers onto a conveyor and then onto a vibrating screen to remove oversized items. The residue travels to another screen that separates particles into two components, less than and greater than 2 inches in diameter. The particles greater than 2 inches are sent through an electromagnetic drum separator to lift out the ferrous material. The particles smaller than 2 inches are sent through a wet-screen separator for ash removal and to an impact mill that crushes the material to enhance the recovery of non-ferrous material in the next stage of separation.

Next the two streams are sent to a heavy medium separation stage where the heavy nonferrous material is separated from the total nonferrous stream through a process involving flotation. At the end of this stage, it is straightforward to segregate the various constituents: (1) the large oversized objects for placement in landfills; (2) ferrous items greater than 2 inches in size; (3) ferrous items less than 2 inches in size; (4) large aluminum pieces and aluminum flakes; and (5) graded residues of other makeup.[30]

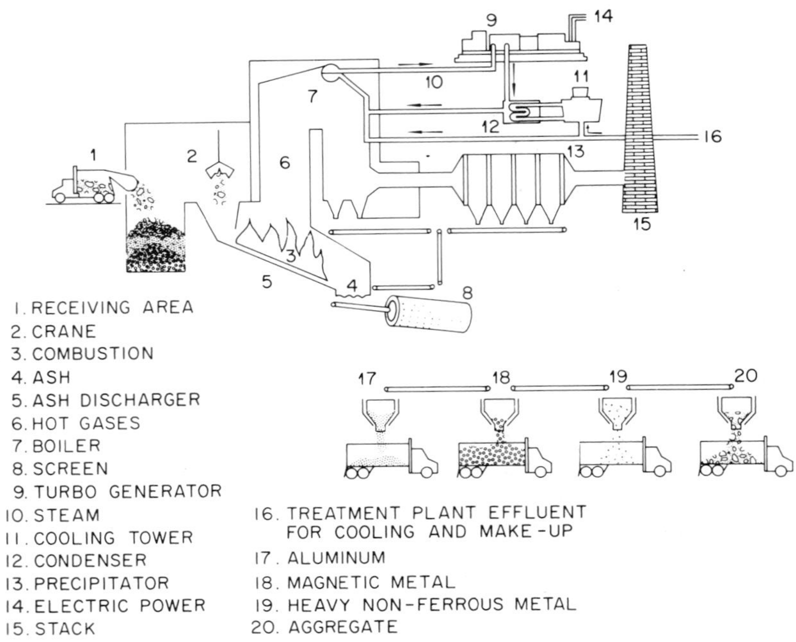


Figure 6-15 Resource recovery facility fly ash separation technique

## 6-14 OTHER RESOURCE RECOVERY TECHNOLOGIES

The importance of developing the most effective technologies for the disposal of solid waste is clear. Interesting work is underway on systems that employ two-stage modular combustion units and fluidized bed combustion units. As discussed in Chapter 8, research of this type should be encouraged and supported.

Historically, there have been other approaches to solid waste disposal based on the resource recovery concept. They have never been as popular as mass incineration and, indeed, most have failed. Nevertheless, it is important for the reader to be aware of the existence of alternate processes. Such alternatives fall into the category of either chemical or biological processes. An example of each of these is reviewed below. Pyrolysis is a disposal method based on destructive chemical distillation. Composting is a method based on the biological decomposition of solid waste by microorganisms.

### Pyrolysis

In incineration heat is generated through the combustion of refuse. In pyrolysis heat is supplied in the absence of air to cause the release of gases which may provide energy. An example of a pyrolytic reaction leading to the conversion of the primary constituent of paper products is given by the reduction of cellulose:

$$C_6H_{10}O_5 \dashrightarrow CH_4 + 2CO + 3H_2O + 3C$$

In the presence of heat, but not oxygen, the molecule of cellulose changes into a molecule of methane ($CH_4$), two carbon monoxide molecules (2CO), three water molecules ($3H_2O$) and three carbon atoms (3C). The methane and carbon monoxide are gases which can be recovered and burned. The carbon residue remains as a furnace residue which, depending upon other wastes treated in the pyrolytic chamber, contains various metals, oxides and minerals as well.

For a variety of reasons the pyrolytic process has not proven commercially viable. In the past 10 years 4 pyrolysis plants operated in the United States. At present, to our knowledge, none is operating.

### Composting

A discussion of solid waste disposal and resource recovery would not be complete without reference to composting. In composting, conditions are created to facilitate the breakdown of organic matter through anaerobic and aerobic bacterial action.

If one excludes metal objects, leather, rubber and plastics, most solid waste consists of organic material. If this organic material is acted upon by bacteria in a controlled manner, the end product is a dark brown or black substance called humus.

Clearly, composting requires the separation of the waste stream into organic and inorganic components. The organic material is then placed in either an open field or in a mechanical system where bacterial activity can proceed. Initially, the material is heated to 130° F, or more, by the bacterial breakdown of the most easily decomposed compounds. Numerous types of bacteria participate in the decomposition process; different ones are triggered into action as the temperature changes. Moreover, the different types of bacteria preferentially attack different kinds of organic matter in the solid waste in different ways.

In open-land composting, the organic material is spread into ground furrows and turned once or twice per week for a period of about five weeks. The resulting humus is removed, ground and marketed. In this kind of composting approximately 2.5 acres of land is required to process 50 tons per day. An additional acre of land is required for every additional 50 tons per day capacity. Highly mechanized composting operations improve the efficiency in the use of land, but still require at least half the space of the open field operations.[31]

To facilitate the composting process, solid waste should be ground into pieces no larger than 3 inches in diameter; the compost should be turned on a regular basis to introduce a new supply of air; the moisture level should be kept in the 50 to 60 percent range; and the temperature should remain in the 130 to 140° F range. To control pathogens in the waste stream the temperature should be elevated to the 140 to 158° F range for a period of at least 24 hours. As indicated above, composting is possible utilizing anaerobic and aerobic bacteria, but because the anaerobic process produces significant odors, this technique is rarely employed.[32]

There are several important environmental considerations that must be taken into account in composting. One concern is that in the mechanical shredding operation which is required to prepare the solid waste for composting, it is nearly impossible to prevent some metal from entering the material to be composted. Metal present in the humus can lead to all of the problems associated with heavy metal toxicity.[33]

Since the aerobic composting process requires the solid organic wastes to be exposed to air for an extended period of time, there is the additional problem associated with surface blown litter leaving the composting area. In addition, great care must be taken to minimize odors released in composting.

While it is evident that composting represents one means for recovering resources from solid waste, it clearly is not suitable at the present time for use in large metropolitan areas where the land requirements, and other environmental issues, make the process prohibitively expensive. The authors are not aware of any major municipal composting facilities now in operation.

Figure 7-1 Alternatives for disposing of solid waste.

# CHAPTER 7

## COMPARATIVE ANALYSIS OF PRINCIPAL DISPOSAL METHODS: SELECTING AMONG ALTERNATIVES

### 7-1 INTRODUCTION

The problem facing many communities in the years ahead is to select the best, most appropriate, alternative or combination of alternatives to manage disposal of their municipal solid wastes. The choices available are limited; each has advantages and disadvantages. These vary from one locale to the next. The best strategy for municipality A may not be the best strategy for municipality B, even though the two cities have the same population, the same population density, and produce the same amount of garbage and trash of the same composition. The best choice will depend upon a variety of factors which include the source of drinking water and its location; the cost and availability of land; the volume of wastes requiring disposal; meteorologic conditions; ambient air quality; transportation options; energy costs; and existing and potential markets for recycled materials. In this chapter we compare and contrast, in a generic way, the advantages and disadvantages--real and perceived--of each of the several strategies now available for disposal of garbage and trash, and outline a process a municipality might use in selecting the best strategy or combination of strategies for its particular situation. Ocean disposal is included even though it is not now a viable option in the United States. Its status may change, at least for stabilized ash from resource recovery facilities.

We draw upon information presented in earlier chapters, summarize important points, and present the results in forms which facilitate comparison of the various alternatives. In examining the alternatives we take account of environmental and public health factors, economic factors, laws and regulations, uncertainties and the need for long-term planning.

In this chapter we direct our attention to the disposal of those materials which, for whatever reason, are not recycled. An imaginative, assertive and effective recycling program could significantly reduce the size and complexity of a community's solid waste problem, but can not eliminate it. There are other important reasons to move toward recycling. These are considered in Chapter 8.

## 7-2 LANDFILLS

Landfilling is the oldest and most familiar method of solid waste disposal. It is a flexible waste management strategy which permits acceptance of wide ranges in the amounts, kinds, and compositions of solid wastes. Many of the objections to the uncontrolled town dump can be reduced significantly, and some even eliminated entirely, through proper siting, design and operation of a sanitary landfill. For relatively small isolated communities, landfilling may be the most appropriate choice. This situation may change, however, as resource recovery industries develop small modular units suitable for communities as small as 10,000 people.

Landfills require relatively large amounts of space and even under the best of circumstances are aesthetically undesirable. Because of their unattractiveness to the senses, identifying and securing acceptable sites in densely populated areas is difficult. Resources such as metals and glass are not readily amenable to on-site recovery at landfills. Only one resource--methane--which is generated by decaying organic matter can be recovered effectively on site. Indeed, under certain circumstances, it may have to be recovered to eliminate the potential for explosions and for other safety reasons.

Vectors--birds, insects, rodents and other animals--which visit dumps can carry pathogens back to people and cause serious illness. The problem can be reduced with a daily cover of approximately 6 inches of fresh dirt, but it cannot be eliminated. If suitable space is available, many objections to landfills can be reduced through proper siting, design and operation. Most landfills, however, are not designed and operated properly, and many are improperly sited.

At least in the short term, landfills are the most economical disposal strategy for many communities. It is important, however, to take into account the ultimate costs associated with the preservation of environmental quality in weighing economic considerations. In metropolitan areas the primary cost may be land acquisition, although construction and operating costs can be significant. New York City estimated that in 1984 dollars the capital costs for construction of a 1,000 ton per day sanitary landfill facility was $14,506,000. The cost did not include land acquisition, but did include engineering design, permitting, materials, equipment, installation systems, access roads, maintenance, administration building and earth moving equipment. According to one study, 90 percent of operating costs of a landfill are for fuel; the remaining 10 percent for labor, maintenance, utilities, and miscellaneous expense. In 1984, tipping fees at landfills across the country ranged from $0 to $56 per ton. The 1984 national average landfill fee of $11.93 per ton was 13 percent higher than the 1983 value. The average tipping fee in the northeast was more than twice that for any other region of the country, reflecting the decreasing availability of landfill capacity.[1] Even the highest tipping fees do not reflect the full costs of disposal.

## 7-3 RESOURCE RECOVERY INCINERATION FACILITIES

Resource recovery incineration facilities require less land than landfills, but more than for ocean dumping. They offer greater opportunity for on-site recovery of materials, particularly ferrous metals, than do other disposal strategies. Resource recovery strategies offer opportunities at several levels to turn a waste product into a resource: energy from burning can be converted into steam for heating or for electricity; metals, both ferrous and non-ferrous can be recovered; and the residual ash has a number of potential uses.

Resource recovery facilities decrease significantly, but do not eliminate, the amount of waste products that must be "disposed of." They also change its character. Resource recovery facilities decrease the volume of wastes by about 90 percent and their mass by about 75 percent. The residual product is a combination of fly ash, bottom ash and, in some cases, scrubber waste. In parts of the world the ash is used for roads, and in concrete as a substitute for sand. There are other potential creative uses for the residual waste products.

With a proper research and development effort, it may be possible to develop methods to stabilize the ash into solid blocks which are strong and environmentally safe and can be used in the construction industry. In coastal areas, blocks may be used for construction of shore protection devices--groins, jetties, seawalls, revetments, and gabions--and for construction of artificial fishing reefs. Scientists already have demonstrated that ash from coal-fired electric-generating stations can be stabilized to produce environmentally safe blocks which can be used for creation of artificial fishing reefs in the ocean, in estuaries and in freshwater lakes.

Resource recovery facilities reduce the potential for groundwater contamination relative to landfills, but increase the potential for atmospheric contamination.

In comparing landfilling with the resource recovery option for a given region, one must determine if it is environmentally safer to destroy in a period of a few minutes a one-ton batch of refuse through incineration, with a residue of ash products and chemicals being instantaneously released into the atmosphere, or to have this one-ton mass slowly decompose in a landfill over a period of decades, releasing other chemicals into the atmosphere and particularly into the surface waters and groundwater. Clearly, the answer depends on detailed questions such as the quantity, toxicity and pathways of the chemicals released, and upon site-specific characteristics.

Questions pertinent to the environmental impact of resource recovery facilities center on the extent to which heavy metals, acid gas and toxic organic emissions pose potential health hazards. Numerous studies indicate that the risks from well-designed, well-operated modern resource recovery facilities with modern emission control devices fall within the range now accepted for protection of human health. Nevertheless, because solid waste disposal will be a persistent activity and a growing societal concern, industry and government have an ongoing responsibility to ensure that the applicable technology is continually assessed and upgraded.

Aesthetically, resource recovery facilities are the most attractive of the land-based municipal solid wastes disposal alternatives. Economically they are attractive because the funding for their construction and operation often comes from the private sector and is recovered through tipping fees and the sale of energy and recovered materials. In 1985, the cost of construction of a resource recovery plant capable of processing 1000 tons per day was about $80 million. A 1975 study estimated that processing of municipal solid waste in a resource recovery facility cost between $15 and $32 per ton of waste and that income from sale of energy and recovered materials ranged from $5 to $17 per ton of waste. To offset the difference, tipping fees were charged at a rate of $3 to $21 per ton at plants processing over 1,000 tons of municipal solid waste per day.

## 7-4 OCEAN DISPOSAL

For municipalities close to the ocean, ocean disposal of municipal solid wastes--unprocessed or after different degrees of processing, including incineration--has obvious advantages, at least in the short term. Current legislation does not prohibit the use of the ocean for dumping of municipal solid wastes, but does prohibit the dumping at sea of some common inert components of trash including plastics. The ocean has been a convenient dumping ground for societies throughout the world for hundreds and even thousands of years. Ocean disposal can be inexpensive. It requires little in the way of capital investment; primarily barges and tugs. There is little potential for pollution either of groundwater or of the atmosphere. In general, the pathways for chemical pollutants to return to people and, as a result, to affect public health, are longer and less direct than for landfilling or for resource recovery. This is not always the case, however. Contamination of seafood products could have serious public health implications. The ocean has an enormous, almost infinite, capacity for assimilating certain kinds of wastes. The Not in My Back Yard (NIMBY) syndrome is minimized. But, there are problems.

On the negative side, the ocean is a common resource and any dumping of waste products is considered by many to be an unacceptable threat to its living resources, and to other uses society wishes to make of the ocean. Dumping of unprocessed, unsorted garbage and trash could cause aesthetic problems because floatables could foul beaches and nearshore areas, causing disastrous impacts on recreational activities with large economic losses. Plastics and other litter pose a threat to marine birds, mammals and fish. Rotting garbage consumes oxygen dissolved in the water and in nearshore areas could cause waters to go anoxic and produce massive mortalities of finfish and shellfish. Pathogens--bacteria and viruses--associated with wastes can be taken up by fish and particularly by shellfish, causing serious human illness, if the fish are eaten raw.

In principle, municipal solid wastes could be disposed of in the ocean unsorted and unprocessed, or after varying degrees of packaging and processing. The simplest processing would be to bale the unsorted wastes before disposal. This practice has advantages over disposal of unprocessed municipal solid wastes. It minimizes dispersal of floatables and of low density materials by waves and currents. The bales also could be used to create fishing reefs to increase local secondary productivity. Another approach in processing would be to burn garbage and trash in a resource recovery facility, generate steam for heat or electricity, separate out glass and metals from the ash, stabilize the ash into blocks and use the blocks to construct artificial fishing reefs, offshore islands, or shore protection structures. Unstabilized (loose) ash from resource recovery plants also could be dumped into the ocean. Particular attention would have to be paid to site selection and time and method of discharge to minimize the potential for adverse impacts.

The effects of ocean dumping of loose ash and stabilized ash are largely unknown. A significant research effort would be required to evaluate this option, and changes in legislation would be required if it were to be implemented.

A comparative assessment of the major impacts of each of the three primary disposal alternatives--landfilling, resource recovery and ocean disposal--is presented in Table 7-1.

## Table 7-1 Assessment of impacts of different disposal alternatives

| POTENTIAL EFFECTS | SANITARY LANDFILLING | OCEAN DISPOSAL | RESOURCE RECOVERY |
|---|---|---|---|
| I. *Environmental* | | | |
| **A. Potential for Drinking Water Pollution** | | | |
| 1. Groundwater | 1. Significant; proper siting; and operation of landfill can decrease risk. | 1. Zero | 1. Less than landfilling; volumes of wastes are reduced making containment easier and they are less refractory. |
| 2. Surface water | 2. As above, although risks are less and more easily reduced. | 2. Zero | 2. As above. |
| **B. Potential for Air Pollution** | | | |
| 1. Particulates | 1. Modest. Can be controlled by daily covering. | 1. Near zero; although some bacteria and viruses may be transferred to atmosphere by bubbles in breaking waves. | 1. Very low. With scrubbers and baghouse filters particulates can be reduced to very low levels. |
| 2. Gases | 2. Modest. Methane can be a problem unless controlled; a potential resource. | 2. Zero | 2. Low to Modest. Levels of most gases can be reduced to very low levels with proper combustion and scrubbers. Levels of uncertainty of potential health effects of dioxins and furans are equivocal. |
| 3. Odor | 3. High. Can be reduced with daily burial. | 3. Zero | 3. Near zero. |
| **C. Potential for Aesthetic Pollution** | 1. High; particularly for Visual and Olfactory degradation. Noise from earth-moving machines also can be objectionable. Problems reduced with proper siting and operation, but not eliminated. | 1. High if raw, unsorted, unprocessed wastes are discharged because of floatables which can foul beaches. | 1. Low. Plants can be attractive, nearly noise and odor free. |
| **D. Vectors** | | | |
| 1. Spread of disease by rodents, birds, insects. | 1. High. Can be reduced with daily burial. | 1. Small, although contaminated fish and shellfish can be vectors for pathogens. | 1. Near zero. |

**Table 7-1 (continued)**

| POTENTIAL EFFECTS | SANITARY LANDFILLING | OCEAN DISPOSAL | RESOURCE RECOVERY |
|---|---|---|---|
| II. *Economic* | | | |
| A. **Costs** | | | |
| 1. Land Acquisition | 1. Largest cost. Requires more land than other options. | 1. Zero | 1. Modest. Land required is far less than for landfill. |
| 2. Capital Costs | 2. Modest, but increases for sanitary landfills which require liners and other precautions. | 2. Barges needed | 2. High; about $80 million for for a 1000 ton/day plant. A plant of this size would satisfy the needs of a municipality of about 400,000 people. |
| 3. Maintenance and Operation | 3. Can be significant. Tipping fees at landfills in 1975 ranged from $2 to $10 per ton. | 3. Vessels and fuel | 3. According to a 1975 study processing costs ranged from $15 to $32 per ton. Income from recovered materials and energy ranged from $5 to $17 per ton. Tipping fees ranged from $3 to $21 per ton at plants processing more than 1000 tons per day. |
| B. **Benefits** | | | |
| 1. Gas recovery | 1. Methane recovery is possible. | 1. Zero unless garbage is confined to covered underwater pits. Not demonstrated. | 1. Zero |
| 2. Metals | 2. Recovery on site is not practical. | 2. Zero on site | 1. High |
| III. Uncertainty In Controlling and Predicting Undesirable Impacts | 1. Modest to high Potential for leakage is persistent. | 1. Modest | 1. Low to Modest. The residual uncertainty is primarily in the public health hazards of prolonged exposure to low levels of dioxins and furans. Unlike a landfill, the emissions from a resource recovery plant can be eliminated entirely by shutting down the plant. |
| IV. Other Factors | | | |
| 1. Need for Secondary disposal method. | 1. Zero. Landfilling is the ultimate solution. | 1. Zero | 1. High. The residue (ash) must be disposed of. It amounts to about 25% of the original volume; as low as 15% with recycling. It can be a resource. |

## 7-5 FACTORS TO CONSIDER

In the sections that follow we examine some of the most important factors a municipality should consider before selecting a strategy, or combination of strategies, to deal with its municipal solid waste problems. Each municipality should take as its starting point the development of the most ambitious program of recycling and source reduction that is practical.

### Environmental and Public Health Factors

*The source of drinking water.*

If drinking water is obtained from groundwater or from local surface water, the landfilling option must be examined with particular care. The potential for contamination of groundwater is greater for landfilling of raw garbage and trash than for any other municipal solid waste disposal option. The potential for contamination of groundwater by landfilling of ash from resource recovery facilities is largely unknown, and should be investigated. If the landfill option is selected, siting should be based on an assessment of hydrogeologic characteristics to minimize chances of groundwater and surface water contamination. Once the site has been selected, proper design, operation and monitoring of the sanitary landfill facility are needed to reduce risk even further.

*Effects on Air Quality*

Landfills often are the source of objectionable odors, and wind-blown litter. These nuisances can be controlled by daily covering with fresh dirt. Landfills also can be sources of methane and a variety of other gases to the atmosphere. With appropriate technology, methane gas can be recovered and used as an energy source.

The particulate emissions of resource recovery facilities can be controlled and reduced to acceptable levels with electrostatic precipitators and baghouse filters. Scrubbers can reduce sulfur oxides and hydrochloric acid to levels which meet federal emission guidelines. Emissions of dioxins and furans and other toxic organics from resource recovery facilities can be minimized by ensuring good combustion conditions. Research is being conducted to determine how these gases are formed and what technologies might be even more effective in controlling their generation, or in removing them from the effluent. This is an important area of research. While federal standards do not now exist for emissions of dioxins and furans, they are being developed.

*The potential for transmission of disease by vectors.*

The transfer of pathogens by vectors--insects, birds, rodents and other mammals--has been a persistent and significant problem of landfills. With daily applications of a 6 inch cover of dirt, the problem can be reduced substantially, but cannot be eliminated. The vector problem at resource recovery facilities is far less. Storage of garbage and trash is inside a closed, controlled environment and the residence time in a facility before burning is short. Access to the refuse by birds is eliminated; access by rodents and insects is reduced markedly in comparison with landfills.

The potential vector problem associated with ocean dumping is different in nature and degree than for either landfilling or resource recovery. Transport of pathogens by birds and insects is still possible, but is limited to the time the garbage is in the barges during transit and at the sea surface following dumping. Transport of pathogens by rodents is eliminated once the barges leave the dock. Accumulation of pathogens in animals, particularly shellfish, and transfer to humans who eat contaminated raw shellfish is a potential problem.

### Economic Factors

*The availability and cost of land close to the community.*

Landfilling requires more land than resource recovery facilities; ocean dumping requires none.

*The volumes of wastes that must be dealt with.*

If volumes are well below some threshold (about 100 tons per day) resource recovery is an economically attractive option only if communities unite to form a regional facility.* The question then becomes which community gets the plant, and which the ash landfill. Smaller resource recovery incineration modules are being developed which could bring resource recovery within range of even smaller communities.

*The cost of electricity in the region.*

If electricity costs are high, some reduction in cost may be possible and desirable with a resource recovery facility. If electricity costs are low, the incentive may be small.

*Markets for recyclable materials.*

If markets for recycled materials exist, or can be created, they can generate new revenues. Recycling also reduces the volumes of material that needs to be disposed of, leading to additional savings.

*A population of about 40,000 produces about 100 tons of solid waste per day. These include, for example, effects on the ecology and the aesthetics which might impact the recreation industry.

*Indirect Economic Effects*

The indirect effects on the regional economy of different disposal options also need to be assessed in selecting among options and in implementing specific options.

*The partitioning of costs.*

The largest costs associated with landfills are for land acquisition. Part of this cost can be recovered when the landfill reaches its storage capacity if it is converted to some other use, such as a park or golf course. Operational costs of landfills are modest; most of the operating cost is in fuel for earth-moving vehicles.

The largest costs associated with resource recovery facilities are for capital construction.

## Applicable Laws and Regulations

*Special Laws and Regulations*

The effects of special laws or regulations that may preclude certain options must be evaluated. For example, the New York State Landfill Law of 1983 mandates closure by 1990 of all landfills in Long Island's Nassau and Suffolk Counties for garbage and trash; and restricts landfilling to ash from resource recovery facilities at special regional landfills to be established outside of the deep recharge aquifer area.

*Jurisdictional Questions*

The articulation of town, county, and state jurisdictions and responsibilities needs to be examined. In some states, responsibility for solid waste disposal rests with the county. In other states it rests with the towns.

## Long-Term Planning

*Long-term plans should be made for the management and disposal of solid wastes.*

Plans should include projections of volumes of waste and should be flexible enough to take appropriate advantage of improvements in technology. They should be adapted to new knowledge about environmental and public health effects of different waste management practices. Surprises should be minimized. If landfills are used, projections should be made of remaining capacities and lifetimes, and appropriate plans should be developed to secure additional sites or to develop and implement alternative disposal strategies well before their lifetimes run out. Public perception and public acceptance must be weighed carefully in developing long-term plans. In any long-term plan, appropriate consideration should be give to reducing the volume of wastes through conservation and recycling, and to developing creative uses for the wastes that remain.

Figure 8-1 Researcher working on solid waste disposal problems.

# CHAPTER 8

## MAKING MOLEHILLS OUT OF MOUNTAINS

### 8-1 INTRODUCTION

The object of any municipal solid waste management strategy plan is to limit impacts on human health and the environment at acceptable cost--or to turn the mountains of garbage and trash into molehills. Such a feat is difficult. Once garbage has been generated and loaded onto collection vehicles, the options for getting rid of it are few. Today garbage is either dumped in landfills or burned. As discussed in earlier chapters, each approach has constraints. The old adage of "the cat comes back" applies to some extent to each process. For landfilling, the principal return path is through the water supply. For incineration, the principal path is through the air. With properly designed disposal systems the impacts on human health and on the environment can be reduced to levels acceptable to most people; improperly designed systems can, however, pose clear health hazards. Even with the best strategy the cat does come back.

To facilitate the making of small molehills, one should first focus on restricting the size of the mountains. To this end we discuss below in some detail the issues associated with recycling and source reduction. Even with extraordinary source reduction and recycling efforts, there will be a burgeoning wasteload that still must be disposed of. The need for continued research on ways to improve landfills and resource recovery facilities is clear. Moreover, though ocean dumping is not now employed as a method of solid waste disposal, there are sound reasons to continue to address the fundamental questions regarding the ability of the oceans to accept solid waste and incineration residue. In this chapter we review several research goals that should be part of future investigations in these areas.

Source reduction includes all strategies that decrease the amount of garbage and trash we produce. Recycling includes all strategies that recover items from the wastestream, and recycle them for reuse. The best long-term waste management strategy would incorporate source reduction and recycling, and there are those who believe that an aggressive program could reduce the nation's municipal solid waste problem from a rampaging river to a "trickle."[1] The arguments for source reduction and recycling are compelling. They reduce the magnitude of the solid waste disposal problem; reduce litter; reduce the demand for energy; reduce pollution of the land, air and water; and conserve scarce raw materials. In spite of these impressive advantages, only about 25 percent of the world's paper, aluminum and steel are recovered for reuse. Why?

Why haven't source reduction and recycling efforts worked in the United States? The reasons are many and complex, but there is one element common to most: the lack of direct economic incentives. The costs of solid waste disposal usually are paid by general taxes and not directly by the people who create the waste products. The costs are enormous. In 1978 Americans spent about $4 billion to collect and dispose of municipal solid wastes. On Long Island, New York, and in many United States cities, expenditures on waste management are second only to those on education.

Source reduction refers to any change in production or consumption practices which leads to a reduction in the solid waste stream. Usually a reduction is thought of in terms of a decrease in the mass, or weight, of garbage and trash that requires disposal. However, home trash compaction units can be thought of as a source reduction technique if disposal is in some way limited by the volume, rather than mass, of waste generated. A reduction in the mass of waste products usually results from decreasing the use of a particular product, from lengthening the useful lifetime of a product, or from decreasing the amount of waste produced during manufacture of a product. Also, when a product ceases to serve its original purpose, the component parts can be recycled for reuse in either the same way, or in another way, to further reduce the waste streams.

## 8-2 SOURCE REDUCTION

Considerable attention has been given to the reduction of hazardous wastes generated during product manufacture. So-called Pollution Payoff Programs have been implemented in some states to educate the producers of hazardous wastes regarding the economic merits of source reduction.[2,3] While no such organized effort has been implemented to reduce the flow of nonhazardous wastes, on-site recycling of scrap materials that result from product manufacture has become widely accepted as an economically sound source reduction technique. Onsite recycling directly back into product manufacture in this manner reduces the potential waste stream by about 5 percent.[4]

There are substantial barriers, real and imagined, to source reduction through decreased use of packaging materials. But such practices could result in a substantial reduction in the volume of solid wastes in the United States. According to the Worldwatch Institute in 1978, packaging accounted for between 30 and 40 percent of all municipal solid waste in the United States[5] (see Fig. 8-2 ). Approximately 75 percent of all glass produced, 40 percent of all paper, 29 percent of all plastic, 14 percent of all aluminum and 8 percent of all steel were used in packaging.[6] The amounts today are probably even greater. Because of health, safety, and particularly because of a myriad of economic factors associated with modern packaging practices, it would be difficult to convince consumers and producers to alter present practices to achieve source reduction. The only arguments that are likely to work are strong economic incentives and disincentives.

"Progress" may represent the major limiting factor that must be dealt with in planning any organized source reduction program involving consumer products. Products often are discarded simply because they have become "obsolete" and are replaced by a more modern version. "Yard" and "garage" sales represent effective means of increasing product lifetimes. In these cases, products that have ceased to serve the purposes of the original owner are passed on to others who may use a product several more years before finally discarding it.

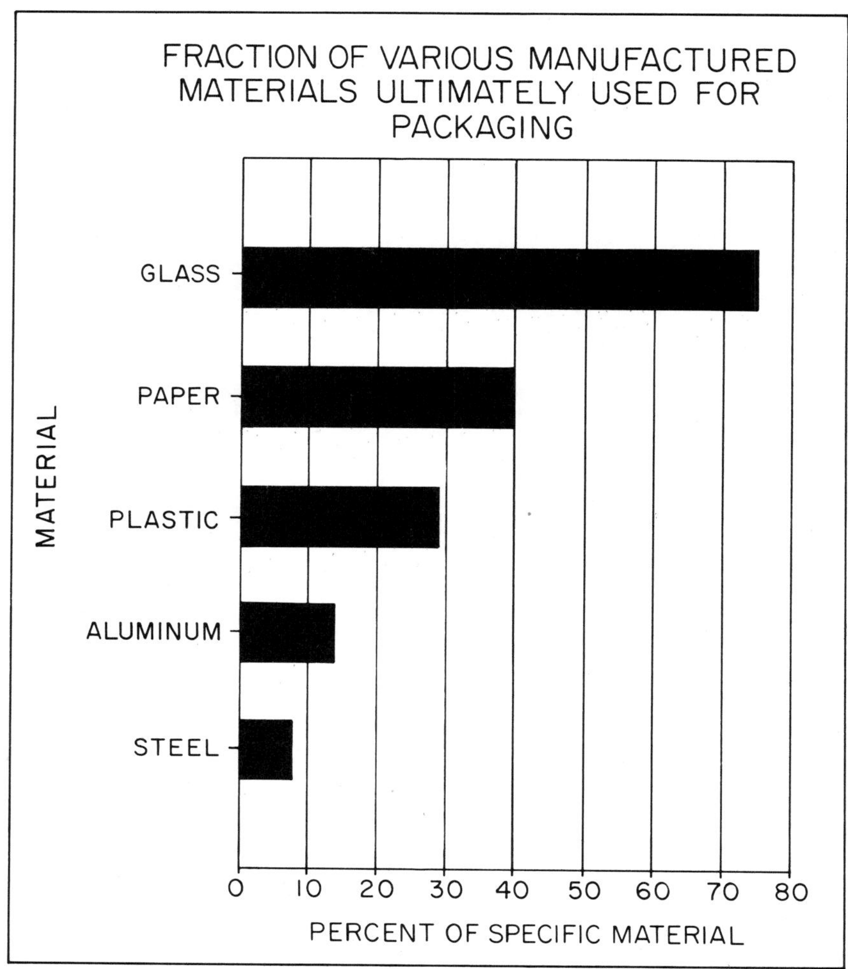


Figure 8-2 Fractions of various manufactured materials used for packaging.

Increasing the durability of products increases their useful lifetimes. In addition to durability, factors such as preconceived expectations for a product's useful lifetime, satisfaction with the performance of a product, its sentimental value, and the original price of the product all affect the consumers' decisions of whether to discard it, repair it, store it, or sell it.[7] Thus, increasing the durability of a product may have only a small effect on increasing the lifetime that a product is used. Additional effort is necessary to define, develop and implement source reduction mechanisms which effectively reduce the rate of wastes generation in the United States.

## 8-3 RECYCLING

At one time, recycling was thought to be the single solution to most wastes disposal problems. The harsh realities of supply and demand, however, have caused recycling to take on a role that currently is of much less importance for reducing the flow of solid wastes than was originally envisioned (see Reference 8 for a review). Recycling rates for a number of important materials are summarized in Table 8 - 1.

**Table 8-1 Recycling Rates for Selected Materials. [Ref. 28 in Chapter 4]**

| Available Material | Short Tons For Recycling | Short Tons Recycled | Percent Recycled |
|---|---|---|---|
| Aluminum | 2,215,000 | 1,056,000 | 48 |
| Copper | 2,456,000 | 1,489,000 | 61 |
| Lead | 1,406,000 | 585,000 | 42 |
| Zinc | 1,271,000 | 182,000 | 14 |
| Nickel | 106,000 | 42,100 | 40 |
| Steel | 141,000,000 | 36,700,000 | 26 |
| Stainless Steel | 429,000 | 378,000 | 88 |
| Precious Metals | 105,000,000 (troy ounces) | 79,000,000 (troy ounces) | 75 |
| Paper | 46,800,000 | 11,400,000 | 19 |
| Textiles | 4,700,000 | 800,000 | 17 |

Recycling can be accomplished through the use of beverage container deposit laws, community recycling programs, or resource recovery operations associated with refuse-derived fuels (RDF) or mass-burning incineration, pyrolysis and composting. The term "recyclable" usually refers to waste components that fall into the categories of ferrous metals, non-ferrous metals, paper, and glass. Energy obtained from refuse-derived fuels (RDF) and mass-burning incinerators is an example where the combustible fraction of waste is the recyclable material, and energy generation is a result of the recycling operation.

The components common to all recycling programs include: (l) a supplier of recyclables, (2) a dealer who collects and disperses the recyclables, and (3) a buyer (see Reference 9 for a detailed discussion of these components). Most successful recycling programs must deal with the problems of generating and sustaining a sufficient supply and finding sufficient demand for the recycled materials to justify the existence of the program on economic grounds. In considerations of economic aspects of recycling, the reduction in waste flow to disposal sites must be included. Waste flow reduction will lower the operating costs and, in the case of landfilling, prolong the lifetime of the disposal site. Also, federal, and in some cases, state, county or city financial assistance can be obtained for recycling operations.

State bottle deposit laws represent an example of how a recycling program can successfully deal with the problems of supply and demand. In this case, beverage consumers are the suppliers of recyclables; the store that sells the beverage containers and accepts returns is the dealer; and the container manufacturers are the buyers of the recyclables. Supply of recyclable materials (the containers) is virtually guaranteed by use of a monetary incentive--the container deposit. Demand for the recycled containers is guaranteed by state laws governing container manufacturers. The success documented for this type of recycling program suggests that it might be adapted readily for recycling other types of products. The New York State bottle law is expected to eventually reduce the flow of solid wastes in New York by about 5.5 percent of the total. State-wide the rate of returns is about 75 percent. In other states it averages more than 90 percent, Table 8-2. As a result of its bottle law, Maine has cut its litter collection costs by 60 percent, and Michigan's economy has gained a net total of 4,600 jobs.

For community recycling programs, the keys to ensuring sufficient supply of recyclables are that the recycling operation involve: (l) minimal inconvenience to customers, (2) incentives for participation, and (3) systems for informing the public about the existence and benefits of the program (see Reference 10 for a case study). In the case of resource recovery operations involving refuse-derived fuels and mass incineration, pyrolysis and commericial composting, the supply of recyclables is determined by the volume of wastes brought to the facility for processing. In most cases, the volume of wastes is determined by the total costs of operating the facility and is controlled through contractual arrangements between facility owners and the municipality involved if the municipality is not the owner.[11]

Source reduction and recycling programs are not incompatible with technological approaches to municipal solid waste management, such as the operation of mass burn resource recovery facilities. Indeed, the two strategies can be complementary. The removal of incombustible materials, such as metals and glass, improves the quality of the fuel and decreases physical abrasion of the furnace and other components of the facility. With proper planning, even the removal of selected combustible materials can enhance the quality of the fuel and improve combustion. If the full benefits of source reduction and recycling are to be realized, they need to be considered from the outset of the planning process to determine the optimum burning capacity of a resource recovery facility. If this is not done, an aggressive paper recycling program conceivably could reduce the supply of fuel to a resource recovery facility designed for a certain volume of waste to a level where the lost revenues from declining energy productions would more than offset the economic benefits of recycling.

**Table 8-2. States with Beverage Container Deposit Laws and Their Container Return Rates [Ref. 5]**

| State | Date Law Effective | Return Rate(%) Bottles | Cans |
|---|---|---|---|
| Oregon | 1972 | 95 | 92 |
| Vermont | 1973 | 93 | 90 |
| Maine | 1978 | 93 | 93 |
| Michigan | 1979 | 96 | 96 |
| Iowa | 1979 | 96 | 90 |
| Conn. | 1980 | -- | -- |
| Mass. | 1983 | -- | -- |
| Delaware | 1983 | -- | -- |
| New York | 1983 | -- | -- |

There is another potential problem that can affect the flow of materials to a resource recovery facility and therefore the amounts of materials available for incineration and recycling. If landfilling is a viable alternative to resource recovery in a particular area, tipping fees at the resource recovery facility must be competitive with those at the landfill. If they are not, the resource recovery operation may fail because of a low supply of materials.[12]

Community recycling programs usually require that participants bear at least some of the burden of materials separation. They may also require that participants transport materials to a collection center. In cases where the collection center is a transfer station, materials separation may or may not be accomplished by station operators.[13,14] In "buy-back" operations, the inconvenience of the requirement that participants transport recyclables to recycling centers is partially offset by a cash incentive. Because curbside collection involves the least inconvenience to customers, many communities are turning to this type of operation to increase participation in recycling programs.[15,16,17] Participant inconvenience can be further reduced by offering back-of-the-yard collection or by providing each customer with separate containers specifically for recyclables.[18,19]

Incentives for participation in community recycling programs are usually monetary. If the recycling program is mandatory, there are monetary penalties for non-participation.[20,21] Seattle, Washington, encourages recycling by using a curbside refuse collection system in which customers are charged according to the number of refuse containers set out for collection.[22] In other cases, customers may be either paid directly for recyclables, as mentioned previously, or given a reduction in refuse collection costs for participation in recycling programs.

The dealer of recyclables usually is an organization that either collects all types of materials, or is one that specializes in one, or several, specific components. This organization may or may not be the same as the one that provides the collection service to individuals. Dealers operate on local, regional, or nationwide bases. They must decide, based upon the markets available and labor costs, the degree of purity--of separation--required for acceptance of materials supplied for recycling. Although low materials purity--a low degree of separation--minimizes inconvenience to participants in a recycling program, high product purity usually is necessary for maximum market value. Thus, some further separation of materials (e.g., removing paper labels from glass bottles or separating aluminum and steel cans) usually is required of dealers or the organization providing the collection service to ensure success of the recycling program. Labor and equipment costs for materials separation play a major role in determining the economic viability of community recycling programs.

Dealers must be prepared to stockpile materials during periods when market demand is low. Considering the fluctuating nature of markets, it probably is best to use dealers who operate on a regional or nationwide basis and who work with a wide range of materials. These dealers have the greatest degree of flexibility. They can compare prices offered by many different companies for a particular recyclable and select the highest bidder. Dealing with a range of materials also has advantages: low demand for one type of material may be offset at any time by high demand for another.

Although considerable emphasis in the past has been placed upon optimization of the supply side of recycling, it is probably the demand side that will determine the ultimate success of recycling for reducing the flow of wastes requiring disposal. Adjusting the collection systems to increase participation by individuals in recycling operations is simple in comparison to the adjustments that will be necessary to increase the demand for many recyclables.

"Technological progress" can be an impediment to recycling. For example, although a particular metal alloy may, at one time, be used for the manufacture of a product, further refinements in the product may rapidly cause that alloy to become obsolete. As a result, once the old product is discarded, there may be no immediate use for the metal alloy parts. Because separation of different metals is usually very time-consuming and expensive, there may be no economic gain from acquiring the discarded metal alloy. There may, therefore, be no demand for that alloy, no matter how large the supply.

In some cases, it is clear that the industrial sector has not put recycling into proper perspective in decisions regarding future product composition. Progress in a given area of manufacturing may have only a minimal impact on the quality of a product, while having a substantial impact on the use of recyclables. For example, pulp egg cartons and meat packaging require only a very low grade of paper, and are ideal candidates for use of recycled paper. These paper packages are being replaced gradually by polymeric foam containers. The improvement in the quality of packaging is minimal, it eliminates a market for recycled paper, and the new containers are not easily decomposed.

Another impediment to recycling of materials is that the quality of products made from recycled components may be lower than those made from virgin materials. This may be due, in part, to the presence of contaminants in the recyclable material. In some cases contamination can be reduced by use of more efficient materials separation methods; in other cases no presently available separation procedure will decontaminate the recycled components. For example, before newspaper can be reused, it must be de-inked. The de-inking process destroys about 10 perent of the original paper and reduces the quality of the final product. Recycled paper is, in general, of lower quality than that made from wood pulping, because the grinding process used in recycling shortens paper fibers. Low-grade recycled paper has received widespread use, for example, as insulation material that replaces more expensive fiberglass and in a range of other applications where a loss in strength is unimportant.

Tin cans consist of a mixture of iron and tin with lead seams, which must be melted down before reuse. Once a melt has been formed, it is virtually impossible to separate the tin from the iron, which substantially reduces the usefulness of the solidified iron product. The lead also attacks and destroys furnace linings. Such examples point out some of the impediments associated with recycling that will be difficult to overcome in the future without major breakthroughs in materials (particularly metals) separation processes.

Logistical considerations also enter into the demand side of recycling operations. Paper mills, for example, usually are located in rural areas near forests, to minimize transportation costs involved with wood pulping. Because recycled paper is generated predominantly in urban areas, transportation costs to bring that paper to the mills are usually much higher. This is one of the principal reasons given by the paper industry for viewing recycled paper as an extra source, to be tapped only during periods of very high paper demand, rather than the principal source of new paper products.[23]

Although about 25 percent of the total waste generated in the United States potentially is recyclable, recycling rates as high as 10 percent of the waste generation rate are achieved only rarely.[24] While community participation in recycling in this country is low, it is likely that demand for recycled products determines the upper limit of recycling rates. Therefore, future actions to increase recycling rates should focus on providing incentives for cooperation of industries. New attitudes toward recycling will be necessary as well as new technologies to reduce problems associated with separation of recyclables from "contaminants". Paper recycling rates could, for example, be substantially increased by bringing paper mills closer to the areas of waste paper generation and by improving paper de-inking procedures. At present, about 10 million tons of waste newspaper are generated each year in the United States, while only three million tons are recycled. In 1985 the apparent consumption rate* of all paper and paper-board products by the United States was 75 million tons. Of this total, about 20 million tons, or 27 percent, were recycled.[25]

---

*The apparent consumption rate of the United States is defined as the total amount of paper produced by the U.S. plus what it imports minus what it exports.

If recycling in the United States is to be successful on a significant level, several things must happen. A successful program will have its roots in economic incentives and disincentives and not in environmental awareness. Still, it is important to embark upon an ambitious and comprehensive educational program at all levels. This educational program should objectively inform citizens of the availability of natural resources and the true costs of present manufacturing and waste disposal practices. Consumers must be made to pay the full costs of the materials they use, including the costs of disposal. Markets must be created for recycled materials. Recycling must be legitimized and facilitated by providing citizens with a variety of opportunities for recycling and by encouraging their use with appropriate incentives and disincentives.

## 8-4 IMPROVEMENTS IN DISPOSAL TECHNOLOGIES

### Landfills

The landfill method of waste disposal has been around a long time, and major breakthroughs in the practice would not normally be expected. Nevertheless, it is clear that the application of sound engineering priciples can help avoid some of the drastic problems caused by poor siting and inadequate containment of leachates.

Research is needed to develop better, more durable landfill liners and caps and to develop more effective ways to capture methane. Research also is needed to develop methods and instruments for rapid, economical and accurate analyses of water samples for contaminants of concern which escape in leachates. Additional information is needed on biological and chemical transformations of substances in landfills and in groundwater, as well as the toxicity of contaminants both individually and in different combinations. Research also is needed to develop more effective and economical methods for the treatment of leachate to remove, or neutralize, dangerous contaminants.

If landfills are to receive ash from resource recovery facilities, research is needed to characterize this material and to establish its behavior under a variety of environmental conditions. This topic requires immediate attention.

### Resource Recovery Facilities

Although resource recovery facilities have been in use for several decades, advances in monitoring sensitivities and incineration chemistry suggest that continued research will be of great value. At present there is no evidence that properly designed resource recovery facilities pose human health hazards, but further research is required both to answer unresolved question and to increase confidence in the safety of these plants. Additional information is needed on the emissions, as well as their effects on the environment and particularly on public health. There appears to be a substantial degree of uncertainty as to (1) the specific locations within resource recovery facilities where furans and dioxins and related compounds are generated; (2) the conditions which promote their formation; (3) their precursors; and (4) the extent to which they are generated *de novo*. There also are few detailed published data on the emission levels of these substances and how they vary with operating conditions, particularly temperature, within a plant, and with composition of the raw refuse. Some data that do exist appear to contradict the common belief that dioxin production decreases with increasing temperature of combustion. There are substantial research efforts underway in the United States, Canada, and Europe which may resolve some of these questions.

More information is needed on the locations and strengths of other sources of dioxins and furans to the environment. We also need a better understanding of the fates and effects of dioxins and furans in the environment and on public health.

Resource recovery facilities exist; they have existed for more than 25 years in Europe and more than 15 years in the United States. Millions of dollars have been spent on research, and significant progress has been made to characterize the combustion process and its products. Additional research is needed to resolve some questions regarding public health and the environment. The questions are complex and can only be addressed through properly designed studies. An understanding of the processes and conditions that control the formation of furans, dioxins and other potentially dangerous substances, such as toxic metals, represents the first step in developing strategies and technologies either to eliminate their generation or to remove them from the effluent. If removed from the effluent, methods must be developed for destruction, or safe disposal.

A small number of carefully designed, comprehensive research and monitoring programs at a few plants would serve our present needs for information far better than a large number of individual studies at a large number of plants. To date, the choice of plants has been limited because of the reluctance of operators to subject their plants to unnecessary scrutiny. A combination of field and laboratory studies might be most effective. The New York State Energy Research and Development Authority has plans to build a small prototype resource recovery facility for experimental use and for the training of plant operators.

Since the results of the research are important to a number of groups of stakeholders--including the public which might be affected and the resource recovery industry-- some form of consortial funding should be developed to support the research. Appropriate partners would include builders and operators of resource recovery facilities; governmental agencies responsible for public health, environmental management and protection; and governmental units at a variety of levels which anticipate construction of resource recovery facilities--regardless of whether or not the research will be done at a facility in that region, or even that state.

Control of the design and execution of these research programs, and the dissemination of the results should be under the supervison of a research committee that is as free as possible of politics and pressure from special interest groups. There is some effort underway to accomplish this. One example is the program initiated by the New York State Energy Research and Development Authority at the VICON Corporation's resource recovery facility in Pittsfield, Massachusetts. This program is laudable, but may need to be expanded and refined. There is great need at present to obtain broad consensus among the experts concerning the important research themes, the methods for translating them into tractable questions, and the kinds of experiments required to answer these questions with acceptable precision and accuracy. These steps should be taken before extensive observational programs are initiated. Research should not be limited to gaseous emissions. The ash has received far too little attention.

Assessment of potential effects on public health requires extensive observations over several decades before patterns can be established unequivocally. Further reductions in possible risks of exposure to furans and dioxins can be achieved immediately by insisting that the best plant designs are used for resource recovery facilities, and that they operate to design specifications. Further refinements of incineration technology, emission control technology and diagnostic monitoring techniques may be achievable within a few years with a proper research and development effort. Current work on two-stage incineration and fluidized bed combustion projects suggest that these are examples of areas where further progress might be expected.

## Ocean Disposal

As a result of local pressures, as well as national and international regulations on the ocean dumping of certain inert materials, the dumping of garbage and trash in the ocean by the United States has ceased, except for that discarded from commercial, military and recreational vessels. However, the ocean may still provide an acceptable depository for some municipal solid wastes, or municipal solid waste residues (ash). Feasibility can be established only after thorough studies of ocean disposal in relation to the full range of other options available.

The effects of ocean dumping may be direct or indirect, immediate or manifested only over the long term. The pathways of waste materials in the marine ecosystem as well as the origin and ultimate fate of pollutants need more study. More information is needed on the persistence of contaminants, and the assimilative capacity of different segments of the ocean for different kinds of waste products and for different disposal strategies. Basic research is needed on the degradation rates and degradation products of different types of organic matter. Lethal and sublethal toxicity and long-term effects of toxic material on marine life should be studied.

Oceanic systems are dynamic and complex. The factors which determine the effects of any particular contaminant on the structure and function of a marine ecosystem are poorly understood. The thresholds between increasing the productivity of an area and altering its species composition are just beginning to be understood. The more subtle long-term effects of waste disposal cannot be predicted at the present time with acceptable accuracy. More basic research is needed on biological ocean dynamics on both large and small spatial and temporal scales, and on how various societal inputs affect these natural processes. Emphasis should be on estuaries and continental shelf waters because these are the areas which receive most of society's wastes.

Research on stabilized and unstabilized ash from resource recovery facilities is just beginning and should be expanded. Questions concerning the products of leaching and their ecological effects, and the stability of both the ash and leachates should be studied under different environmental conditions both in the laboratory and in the field. Information concerning public health risk should be gathered for the entire suite of contaminants associated with ocean disposal of different forms of municipal wastes. Pathogen pathways should be studied. Effective methods of measuring dangers to public health associated with consumption of tainted finfish and shellfish are needed.

The incineration of garbage produces huge amounts of fly ash and bottom ash which must be disposed of. The problem of how to accomplish this disposal will be particularly acute in the metropolitan New York City area. More than eleven million tons of solid wastes are collected annually. This translates into a potential of more than 2.2 million tons of incinerator ash each year; enough ash each year to make more than 65 million blocks, each the size of a standard cinder block.

In light of the problems of groundwater contamination associated with landfilling of incineration ash on Long Island, the New York State Legislature, in concert with local municipalities, recently funded the first study to evaluate disposal of incineration wastes in the marine environment. Fly ash and bottom ash generated by resource recovery facilities may prove to be suitable material for stabilization and marine disposal, but this can be determined only after extensive testing in the laboratory and in the field.

The process of combining ash with lime and water to form a concrete-like material is neither difficult nor expensive and was employed during the Roman era. Volcanic ash from Mount Etna was used to develop superior strength, water-resistant cements for Roman constructions. Some still survive today after 2,000 years. Scientists at the State University of New York at Stony Brook have used the ancient methods of the Romans to develop modern techniques for processing and fixing industrial and municipal wastes into concrete-like blocks. Since 1977 studies have been made of the stability and environmental effects of stabilized blocks made from coal combustion wastes. These have been used to construct artificial fishing reefs in fresh, brackish and full sea water. It appears likely that incineration ash can be used for this and other constructive purposes, but additional research is needed before this can be done with confidence. The feasibility of combining municipal and industrial waste streams to produce creative and environmentally acceptable by-products needs to be evaluated. An assessment of the biological and chemical interactions of different combinations of stabilized waste materials with each other and with the environment has the potential of leading to the identification of acceptable alternatives to landfilling.

Ocean dumping has both ecological and sociological effects. Effects which may be ecologically beneficial may not be socially acceptable. For example, if raw garbage is dumped into an area of low productivity to increase nutrient levels and to stimulate biological production for fisheries or commercial aquaculture, bathers and recreational boaters may not regard the action favorably. One must make the difficult determination of what ecological effects are deleterious and the extent of ecosystem change which is acceptable or desirable.

National and international diagnostic monitoring systems should be developed and implemented. It is important to detect pollution early and trace it to its source.

## 8-5 CONCLUSION

We need to improve our ability to distinguish among the advantages and disadvantages of alternative strategies for dealing with garbage and trash. Only then can we select the most appropriate strategy, or combination of strategies; those which have predictable and acceptable impacts on public health, on the environment and on the economy. This knowledge must be based upon a rigorous assessment of the advantages and disadvantages of each management strategy for a particular region. The best choice will vary from one region to the next even if population, population density, and volume and composition of garbage and trash remain the same.

Can we turn mountains into molehills? Perhaps not molehills, but we can reduce the size and complexity of the mountains of garbage and trash we must deal with if society vigorously pursues recycling and source reduction and insists on the highest standards in its solid waste disposal methods.

# APPENDIX I

## Survey of Methods Currently in Use and Future Plans for Selected Metropolitan Areas

---

This appendix summarizes the approaches used by various cities in the United States in dealing with their problems of municipal solid waste diposal. As discussed in the text, the source of drinking water, population projections, waste generation rates, degree of urbanization, and availability of landfill space all are critical elements in determining the viable options for waste disposal. These are discussed in the context of the current practices and future plans of cities which have responded to a survey conducted by the Marine Sciences Research Center at the State University of New York at Stony Brook during 1985 and 1986.

For purposes of this analysis, the United States is divided into distinct geographic regions to facilitate review of general trends. Plans of specific cities within each region provide more detailed insight into the pertinent issues.

---

### A1-1 NEW ENGLAND REGION

The New England region is made up of six states: Maine, New Hampshire, Vermont, Massachusetts, Rhode Island, and Connecticut.

#### Population Forecast

According to projected forecasts the population in the New England region will increase 13.0% in the 30-year period from 1965-1995. This is substantially below the national growth rate of 34.8% projected for the same 30-year period.

#### Solid Waste Generation Forecast

From 1975-1995 the amount of solid wastes produced in this region is predicted to increase from 9.2 million tons per year to 14.6 million tons per year. This marks a 58% increase in solid waste production. This increase is comparable to the projected increase for the nation as a whole in that same time period.

#### Potential for Resource Recovery

In the New England region about 52% of all the solid waste generated is incinerated. This makes the New England region the leader in utilization of incineration practices nationwide. Many more resource recovery projects are planned over the next 5 to 10 years. All of the states except New Hampshire, have used or are using resource recovery. Connecticut is a leader nationwide in feasibility studies and applications of resource recovery technology.

### Methods and Products of Resource Recovery

Mechanical sorting at a resource recovery plant seems to be the preferred method of recycling in the region. Aluminum and glass recovery in the New England region may or may not be economically feasible since there are very few glass and aluminum plants in the region. Some resource recovery plants separate the materials by air separation, magnetic separation, shredding, and screening. Hydropulping will play a major role for paper fiber recovery. This region has the capacity to absorb far more paper fiber than can be produced locally.

## BOSTON (MA)

### Population Forecast

In the 10-year period from 1970-1980 the population of the city of Boston decreased from 641,071 to 562,994. This decrease marked a total decline in population of 12%, or a 1.2% decline per year. This trend is opposite that expected for the New England area during the 30-year period from 1965-1995. In this time the population of the New England region is expected to increase 13.0%, or 0.4% per year.

### Past Practices

In the past, Boston relied on both public and private land-fills, incineration and, since 1975, minimally on a waste-to-energy facility in the town of Saugus. In November 1975 the city's municipal incinerators were closed. In July, 1980 the last city landfill was closed.

### Present Practices

In 1975 Boston began sending approximately 110 TPD (about 20% of the solid wastes produced in the city at that time) to the Saugus waste-to-energy facility. Since July, 1980 approximately 550 TPD (about 80% of the current solid wastes) have been disposed of in privately owned landfills outside of the city. Each of these avenues continues to be exploited as new options for disposal and resource recovery are explored.

### Future Plans

A recent negotiation with the Signal Corporation has been suspended. The plan had two components. The first was to increase the capacity of the Saugus waste-to-energy facility to handle 50% of the City's disposal needs. The second component was to transport the remaining 50% of the City's garbage to the Braintree transfer station for rail transport to an 1800 TPD refuse-derived fuel facility in Rochester, Massachusetts.

With suspension of these negotiations, an alternate in-city plan is being pursued. The American REF-FUEL Corporation would build, own, and operate a $185 million Waste-to-Energy facility. This facility would have a capacity of about 1500 TPD. It would be located in the South Bay area at the site of the old incinerator. This contract would cost $90 million less than trucking wastes out of town and cost the city $177 million over a 20-year period.

### Decision Factors

Boston's two main considerations in choosing a method of solid waste disposal are: 1) public health; and 2) traffic problems. Health concerns relate to dioxins and dibenzofurans. Some researchers argue that these compounds pose an excess cancer risk to residents in the vicinity of proposed resource recovery facilities. Traffic flow can be seriously impaired if new waste collection and drop-off facilities are poorly located.

## BRIDGEPORT (CT)

### Population Forecast

Bridgeport has a population of approximately 143,210. By the year 2000, the population is expected to reach 162,000. This represents a 13% increase over the next 15 years, which is equivalent to the predicted percent rise for the New England region as a whole.

### Solid Waste Generation and Management

The City of Bridgeport currently generates about 110,000 tons of solid waste per year. About 126,000 tons per year, or a 14% increase, is anticipated by the year 2000.

In the past, Bridgeport has both landfilled and incinerated its municipal solid waste. The garbage was landfilled at a 34-acre landfill site. Incineration took place at the Asylum Street Incinerator, which was not equipped with an energy or materials recovery system. Today, both the landfill site and the incinerator are closed. Bridgeport attempted a resource recovery project about 5 years ago. The 1800 TPD facility used mechanical and chemical processing to produce a synthetic fuel product called "ECO-Fuel II". After 17 months of testing, the plant shut down without approaching the design capacity. Environmental and technological problems contributed to its closure.

At present, all of Bridgeport's solid waste is exported from the city to the Shelton landfill site in Hamden, CT. Future waste management plans for Bridgeport include the construction of a waste-to-energy plant which will be built by the CRRA-Signal Rescue Company.

---

## A1-2 MIDDLE ATLANTIC REGION

The Middle Atlantic region is made up of three states: (1) New York, (2) New Jersey, and (3) Pennsylvania.

### Population Forecast

The population of the Middle Atlantic region is expected to experience a 1.4% decline over the period 1965-1995. This trend is opposite the trend projected for the nation for this time period.

### Solid Waste Generation Forecast

The production of solid wastes in the Middle Atlantic region is projected to increase by 51% over the 30-year period from 1965-1995. This marks an increase from 28 million tons per year in 1965 to 43 million tons per year in 1995. This increase in waste production is lower than the projected national solid waste production increase.

### Potential for Resource Recovery

The Middle Atlantic region is a highly industrialized region making it ideal for resource recovery. Within economically feasible distances of solid waste generation centers there are plants to absorb a rich assortment of recovered materials. The industry will have a hard time using all of the recovered glass until the mid-1980's given the projected quantities that will be available. Residual aggregates, bottom ash, will probably also be an unsuccessful resource for recovery. It appears that all other major recoverable materials and energy will have ample markets to support them.

### Practices at Time of Last Major Study (1976)

At the time of a 1976 study, the major methods of disposal in the Middle Atlantic region included landfilling, incineration, and resource-derived fuel facilities. Resource recovery incineration will probably reach its maximum potential in this area because of its dense population, and limited land availability.

## NEW YORK CITY (NY)

### Population Forecast

New York City is made up of five boroughs: (1) Manhattan, (2) Brooklyn, (3) Bronx, (4) Queens, and (5) Staten Island. The population in New York increased from 7.8 million to 8.4 million in the 19-year period from 1965-1984. This marks a population increase of 7.7 percent.

Manhattan is the smallest and most densely populated of the five boroughs. As Manhattan reaches its maximum capactiy, population growth will slow and the excess growth will take place in the other boroughs. Population density and land availability play a role in the potential for solid waste disposal sites. Manhattan appears to be the least viable of the five boroughs for solid waste disposal sites because there is little unoccupied land.

**Table A1-1 Area, population density and population of New York City and each of its five Boroughs**

| Borough | Land Area (sq. miles) | Population (in 1,000s) | Population per sq. mile(1982) |
|---|---|---|---|
| Bronx | 41.7 | 1,163 | 27,890 |
| Brooklyn | 70.2 | 2,241 | 31,923 |
| Manhattan | 22.2 | 1,423 | 64,099 |
| Queens | 108.6 | 1,900 | 17,495 |
| Staten Is. | 58.7 | 360 | 6,733 |
| NEW YORK CITY (total) | 301.5 | 7,086 | 23,502 |

## Solid Waste Generation Forecast

The volumes of solid wastes produced in New York City are expected to increase from 26,550 TPD in 1983 to 27,815 TPD in 1996, an increase of 4.8%. This annual increase of 0.36% is well below the rate of increase projected for the United States as a whole.

## Past Practices

New York City's solid waste disposal methods have changed greatly over the years. The number and variety of alternatives has decreased considerably over time. The 1930s marked a time of diversity for New York's solid waste disposal alternatives. The options included: (1) ocean dumping, (2) recycling, (3) landfilling, (4) island extension, and (5) incineration.

The diverse network of solid waste disposal alternatives present in the 1930's changed for three principal reasons, increased environmental regulations, decreased land availability, and economic feasibility.

Growing environmental concerns sparked the passage of new and stricter regulations governing solid waste disposal. Ocean dumping was banned in 1934. This legislation eliminated the method of disposal that had supplied 12 percent of the city's waste disposal needs.

The ban on uncontrolled incineration of solid wastes in municipally or privately operated incinerators led to the closing of over 17,000 incinerators in the mid 1970s. The overwhelming majority of these incinerators were located in residential dwellings, municipal incinerators numbered twenty-two. Because residential incinerators were not subject to emission standards, the city's lack of control caused great concern about the potential hazards of exposure to harmful emissions.

Active landfilling in wetland areas was restricted as the value of these areas was realized. Wetland landfilling was a method of island enlargement prior to the imposition of the restrictions.

The ultimate means of solid waste disposal, the landfill, was also subjected to tightening regulations. Many of the city's 89 landfills that were in operation in 1934 were closed in the 1970s because they were unable to meet regulations.

The feasibility of each method of solid waste disposal had to be evaluated on the basis of the new regulations and available land resources. Until recently, landfilling appeared to be the most acceptable method of solid waste disposal. However, due to land use constraints, landfilling is no longer considered a feasible option for future disposal of New York's solid wastes.

## Present Practices

New York City relies on three methods of solid waste disposal: (1) landfilling, (2) incineration, and (3) exportation. Despite changes presently being effected, landfilling currently remains the most important.

### Landfilling

As of January 1, 1986, two landfills were operational in New York City: (1) Fresh Kills and (2) Edgemere. These landfills provided approximately 80 percent of the capacity for disposal each day.

The Fresh Kills landfill, located on Staten Island, is the largest of New York's landfills and accepts between 19,000 and 20,000 TPD. Its remaining life expectancy is between 14 and 18 years. The upper limit of 18 years is based on a two-year extension that would result from the 1988 opening of the Brooklyn Navy Yard resource recovery facility.

The Edgemere landfill, located on the Rockaway Peninsula, represents a very minor part of the overall solid waste disposal plan for New York City, handling only 500-750 TPD. This county owned landfill could achieve a life expectancy of 48 years if waste handling techniques are upgraded.

### Incineration

Incineration accounts for the disposal of 2,000 tons of waste per day is New York City. The Southwest Brooklyn and Greenpoint incinerators are the only city incinerators still operational in New York.

### Exportation

Exportation accounts for the disposal of approximately 2,000 TPD and is conducted by the private waste hauling industry. Wastes are sent to neighboring New Jersey for disposal. The New York City Department of Sanitation had explored the possibility of exporting larger quantities but abandoned its efforts when neighboring states initiated stricter land use regulations.

Landfilling, currently the most significant of the three disposal methods, is rapidly diminishing in its capabilities to meet New York's growing solid waste disposal needs. The Fresh Kills facility was recently upgraded to three active components which include truckfills at Muldoon and Victory Avenues and Plants 1 and 2 at the Fresh Kills Marine Operations Facility. Because the Fresh Kills facility now accepts such a large fraction of New York City's total municipal solid wastes:

(1) the system is vulnerable to mechanical breakdowns and disruptions of all sorts,

(2) costs will increase as further upgrading procedures are undertaken to prolong its life

(3) the potentially negative local environmental impact will heighten in intensity as a result of disposal at one site, and

(4) valuable landfill capacity will be rapidly depleted at an ever-accelerating rate.

Proposals for the implementation of resource recovery as an alternative method of solid waste disposal are being developed and reviewed.

**Future Practices**

New York is in the process of selecting and implementing a solid waste disposal system that will minimize adverse environmental, community and economic impacts and be operationally reliable and efficient. The system must have a capacity of 27,000 TPD and provide a network of strategically located disposal facilities. It is being planned and timed so that the city is never short of necessary disposal capacity. New methods and capacity for waste disposal are necessary through a phased implementaion because:

(1) resource recovery facilities can take 6 to 8 years to implement,

(2) waste reduction measures take time to implement,

(3) landfilling is becoming less acceptable environmentally,

(4) total conversion to resource recovery is a large capital commitment,

(5) a crisis will develop if measures are not taken to avert it; the economic risk of waiting is too great,

(6) through phased implementation, state-of-the-art technology can be employed, and

(7) landfilling space can be preserved.

The most viable long range solution for New York City appears to be resource recovery. Some landfill space must remain for the disposal of noncombustibles and ash.

The solid waste disposal plan for New York City combines extension of landfilling leases, resource recovery, recycling, and possible exportation. The City of New York would like a plan capable of accomodating all of the City's garbage for at least 20 years. At present New York is well below this optimum planning time.

New York has undertaken steps for a phased-in development of resource recovery facilities. Over a 12 year period seven plants would be built, in addition to the Brooklyn Navy Yard resource recovery facility. A two phase development strategy was initiated with the closure of the Fountain Avenue and Pennsylvania Avenue landfills. To offset this loss, the handling and storage capacity of Fresh Kills landfill were increased.

Phase One, to be completed by 1990, is intended to reduce the heavy reliance on landfilling with the introduction of three truck-fed resource recovery facilities of approximately 2,000 TPD capacity each. These facilities are spread throughout the city to decrease haul distances. One of these facilities, the proposed Brooklyn Navy Yard resource recovery facility has been approved and Signal Environmental Systems was contracted for its development. Phase Two, calls for the opening of another two truck-fed resource recovery facilities, again reducing New York's dependence on the landfill.

## HEMPSTEAD (NY)

### Population Forecast

In the 10-year period from 1970 to 1980, the population of the Town of Hempstead decreased from 834,719 to 772,590. This is a decrease in population of 7.4% or 0.74% decrease per year. A 1.4% decline in population is expected for the entire Middle Atlantic region during the 30-year period 1965-1995.

### Solid Waste Generation and Management

Hempstead generates approximately 670,000 tons of garbage per year. This amount is expected to remain nearly fixed over the next few years. In the past, Hempstead has landfilled most of their waste. The town attempted a resource recovery project in the late 1970's. A 2000 TPD RDF processing facility that employed hydropulping to produce RDF for dedicated combustion encountered a number of environmental and contractual problems, and was closed in 1979. Presently, about two-thirds of Hempstead's municipal solid waste is landfilled in the Oceanside landfill. The remaining one-third is delivered to a transfer station (Merrick Facility-BFI Transfer Station). The trash is then carted about 90 miles upstate to Goshen, NY, where it is disposed of in a landfill.

**Future Plans**

Hempstead is planning to employ another resource recovery facility to dispose of its solid waste. This action is primarily a result of diminishing landfill space, and legislation which will allow only ash to be landfilled on Long Island beginning in 1990. A mass-burning facility, having a capacity of 2,250 TPD, is anticipated to be in full operation by 1989. Parts of the defunct RDF physical structure and support facilities will be used in the development of the new mass-burn technology facility.

## HARRISBURG (PA)

**Population Forecast**

In the 13-year period from 1970-1983 the population of the Harrisburg metropolitan area (Harrisburg, Lebanon, Carlisle) increased from 510,000 to 564,000. This increase marks a growth of 10.6%, or an annual growth rate of 0.8%. This trend is opposite to that expected for the Middle Atlantic region during the 30-year period from 1965-1995. During this period the population in the Middle Atlantic region is expected to experience a 0.05% annual decline.

**Solid Waste Generation and Management**

Harrisburg currently produces 65 to 70 tons of garbage per day, or about 25,000 tons per year. Nearly 13 years ago, Harrisburg incorporated a resource recovery facility into its waste management program. The Martin mass-burning facility has a capacity of 720 TPD and receives waste from three counties including material from as far away as King of Prussia, PA. The mass-burn waterwall plant generates steam for the Pennsylvania Power and Light Company for district heating, and also for the Bethelem Steel Company. The residual ash is landfilled. Since the resource recovery facility has the capacity to dispose of all the solid waste generated by the city, no MSW is landfilled except for that rejected for incineration.

**Future Plans**

Harrisburg plans to continue to use resource recovery far into the future. In 1986, the city plans to implement a co-generation project which will produce 8 megawatts of electricity.

---

## A1-3 EAST NORTH CENTRAL REGION

The East North Central region is made up of five states: Ohio, Indiana, Illinois, Michigan, and Wisconsin.

### Population Forecast

The population in the East North Central region is expected to increase 10.5% over the 30-year period from 1965-1995. This is far below the projected growth rate for the nation for the same period.

### Solid Waste Generation Forecast

In the 30-year period from 1965-1995 the amount of solid waste in this region is projected to increase 56.9%. This is slightly below the projected national increase for the same period.

### Potential for Resource Recovery

There is excellent potential for resource recovery in this region. The only recoverable product that does not appear to have a good market is the ash.

## AKRON (OH)

### Population Forecast

In the 13-year period from 1970-1983 the population of the Akron metropolitan area decreased from 679,000 to 652,000, a decrease of 4%, or 0.3% per year. The East North Central region as a whole is estimated to have increased at a rate of 3.5% per year over the same period.

### Solid Waste Generation Forecast

The Akron metropolitan area produced more than 1000 tons of solid waste in 1971. No more specific values of solid waste volumes for the present or future are available for the Akron area.

Waste collection in the Akron metropolitan area cost about $2.5 to 3.0 million in 1983. Disposal cost an additional $1.5-$2 million. At that time, sanitary landfills provided disposal sites for the solid wastes. If 1971 policies of landfilling continue until 1991, the additional land required for solid waste disposal would be between five and six square miles. This amount of land is difficult to obtain within the city limits. Added transportation costs to more distant landfills make this alternative less attractive.

**Past Practices**

Landfilling was the preferred method of solid waste disposal in the Akron metropolitan area until 1981. In February, 1969, concern about rapidly diminishing availability of landfill areas and economic and environmental concerns led to the initiation of solid waste reduction studies. The sanitary landfill in the Akron metropolitan area became an increasingly less favorable means of solid waste disposal as volumes increased and land became less readily available within reasonable hauling distances. Public objection to sanitary landfills was great. Locating new landfills was very difficult to do.

Eventually, a recycling energy system was adopted, which included solid waste reduction in combination with steam production in a resource recovery facility. The steam produced in the Recycling Energy System (RES) is supplied to Central Business District, the plant itself, the B.F. Goodrich plant, and the University of Akron. In all, there are 180 customers.

**Alternate Methods of Disposal**

In 1971, complete conversion to resource recovery was not considered economically or commercially feasible in the Akron metropolitan area. The success of such facilities was judged to be dependent on the sale of recovered materials. In 1986, however, resource appears to be an effective way of dealing with the solid waste disposal problem in densely populated urban areas.

A number of consultants selected it as the best method of solid waste reduction for the Akron metropolitan area.

The Recycle Energy System of the Akron metropolitan area was constructed in 1979 with a design capacity of 1,000 TPD. Its objectives are:

(1) to accept any solid wastes with minimal sorting,

(2) to accept bulk wastes,

(3) to achieve maximum volume reduction of waste,

(4) to produce a residue suitable for landfilling,

(5) to be an environmentally sound system with no water or air pollution resulting from operation,

(6) to be aesthetically acceptable in an urban environment,

(7) to be economically viable in terms of manpower, operation, and maintenance,

(8) to maximize recovery of materials and energy, and

(9) to be a self-sufficient operation.

Operation of the Recycle Energy System was delayed for two years because of problems. The plant was well below the operational design capacity of 1000 TPD. It was decided in 1981 that the plant would either be shut down or undergo major alterations. The Tricil Company was chosen to modify the plant. Three goals of the modifications were (1) to reduce design capacity so that it could handle 800 TPD at a cost of $11,450,000, (2) to operate the steam distribution system, initially on natural gas and later, to switch to solid wastes, and (3) to enter a 10-year operating contract. By December 1982 the RES was operating successfully. Following an explosion in the plant in December 1984, the city terminated its contract with Tricil Resources. The facility is now under contract to the Waste Energy Technology Corporation. The RES performed well in a month-long test during which it processed about 600 TPD. The newly modified plant supplies hot water to two warehouse facilities and steam to the central business district, the B.F. Goodrich plant, and the University of Akron. In addition to steam, the ferrous material recovered from this plant will be sold as scrap metal.

## CHICAGO (IL)

### Population Forecast

In the 13-year period from 1970-1983 the population of Chicago increased from 6,093,000 to 6,119,160, a total growth of 0.4%, or 0.03% per year. This growth rate of 0.03% is well below the annual growth projected for the East North Central Region during the 30-year period from 1965-1995.

### Past Practices

Until 1982 incineration and landfilling represented the preferred methods of solid waste disposal in Chicago. In 1980, Chicago's Department of Public Works began looking into alternatives to landfilling. This represented an effort to expand Chicago's resource recovery capabilities. The operation of four bulk incinerators reduces the volumes of solid wastes, and privately owned landfills accept the ash from the incineration process.

Landfills in Chicago are becoming much less cost effective. Between 1978 and 1982 the cost of operation doubled. The available landfills are expected to reach their maximum capacity in about five years and there is a moratorium on extending existing landfills and opening new ones within the city limits. When capacity is reached, will have to be sought in areas outside of the city. Undeveloped land acceptable for landfills within and close to the city is becoming very hard to find.

At one time Chicago maintained and operated four incinerators. The newest and most technologically advanced is the 1600 TPD capacity Northwest Waste-to-Energy Facility. Since 1980 three of the four incinerators have been closed for failure to meet federal air pollution control standards, leaving only the Northwest Waste-to-Energy Facility. This facility incinerates 30%-40% of Chicago's solid wastes. The remainder is landfilled. This represents one of the largest volume resource recovery facilities in the country. The Northwest Waste-to-Energy Facility is designed for the reclamation of energy for steam production. Steam produced in this facility is used: (1) to drive the plant equipment; (2) to provide hot water for usage in the plant, (3) to provide hot water for vehicle washing in a nearby garage.

Steam will also be supplied to the Brach candy factory. An additional 1.5 million tons of solid waste is collected by private carters from high-rise apartments and commercial establishments, etc., and placed in landfills.

**Present Practices**

Because of Chicago's large solid waste production of 1.1 million tons of solid waste per year (a 1982 statistic), attempts are being made to reduce the costs of solid waste disposal. In addition to simple cost reduction, the Chicago Department of Public Works is looking for methods to decrease the city's reliance on landfilling. Chicago began looking for a dependable, cost-effective method of solid waste disposal in 1980. Studies were undertaken to determine the best ways to expand the city's resource recovery facilities.

A number of consulting firms made preliminary recommendations between August, 1980 and May, 1981. Detailed reports including engineering, economic, marketing and financial analyses were prepared between August, 1981 and August, 1982.

The resource recovery facilities' purpose is to reduce the net cost of solid waste disposal. This is done through maximum energy production and material recycling from the solid wastes. Disposal costs will be reduced as a result of: (1) revenues from the sale of energy produced, (2) revenues from the sale of recovered materials, and (3) reduction of land required for landfilling.

Conclusions of the study offer three alternatives for near-term resource recovery facilities:

(1) reopen the Southwest Supplementary Fuel Processing Facility (SSFPF) with a capacity of 1000-1200 TPD;

(2) rebuild the Southwest incinerator as a 900 TPD mass-burning waterwall incinerator to produce steam for use in adjacent industrial stockyards and convert the steam to electricity for sale to Commonwealth Edison;

(3) build a new mass-burning waterwall incinerator withan 850 TPD capacity. Steam reclamation would be sold to the University of Illinois Medical Center.

In addition to these three alternatives, the studies also proposed other actions to alleviate the solid waste disposal problem. They recommended modification and relocation of transfer stations, and construction of a new resource recovery facility in the Calumet region.

**Future Plans**

Chicago plans to rely almost fully on resource recovery facilities in the future. This method of solid waste disposal seems to fit the important environmental and economic needs of the city.

---

## A1-4 WEST NORTH CENTRAL REGION

The West North Central region is made up of seven states: Minnesota, Iowa, Missouri, North Dakota, South Dakota, Nebraska, and Kansas.

### Population Forecast

The population in this region is expected to increase from 15.5 million to 18 million during the 30-year period from 1965-1995. This marks a 16.1% increase. This is substantially below the projected national increase for the same period.

## OMAHA (NB)

### Drinking Water Source

Omaha receives almost all of its drinking water from the Missouri River.

### Population Forecast

In the 30-year period from 1965-1995 the populaton of Omaha is projected to increase from 539,320 to 694,600. This population growth is extrapolated from the known population growth between 1970 and 1983. During this 13-year period, Omaha experienced an annual growth rate of 0.6%. This growth rate is slighlty above the 0.54% projected annual growth rate for the West North Central region during the 30-year period from 1965-1995.

### Past Practices

Omaha has always relied on landfilling as the preferred method of solid waste disposal. This was and still appears to be a good method for this area.

### Present Practices

Omaha relies on publicly and privately owned landfills. The east side of the city is serviced by a privately owned landfill in Fremont, 32 miles northwest of Omaha. Wastes are collected once a week. They are sent to a bailing plant, then transported in closed trucks to the landfill.

Solid wastes from the west side of the city are disposed of in a county landfill. This waste is sent directly to the landfill without bailing.

The paper stream is reduced by 8% as a result of newspaper recycling contracted out to an insulation company. This represents a reduction of approximately 3.3% in the total waste stream by volume. This reduction is based on a total paper volume which accounts for 41% of the total municipal solid wastes. In 1984, 1,593 tons of newspapers were recycled. The volume of recycled newspaper is expected to increase to approximately 20-25% of the total paper stream.

### Future Plans

Landfilling will continue into the foreseeable future. New lands are being sought by the county. The Fremont site was closed on March 22, 1986 when the contract expired and was not renewed. Wastes from the east side of Omaha are being transported to the county landfill, located south of the city. This facility has a life expectancy of approximately one and a half years.

### Decision Factors

Landfilling appears to be a viable method of solid waste disposal for Omaha. Land cost and availability make this an economically feasible option. Environmental concerns are minimal. According to the Mayor's office, with careful siting and proper maintenance the drinking water should not be threatened and the landfills should not be an environmentally negative factor.

## ST. LOUIS (MO)

### Population Forecasts

In the 13-year period from 1970-1983 the population of St. Louis has decreased from 622,000 to 437,000. This is a decrease in population of 29.7%, or 2.3% per year. This rate is significantly below the 0.54% projected annual growth rate for the West North Central region during the 30-year period from 1965-1995.

### Past Practices

St. Louis relied on incineration, ash pits, and landfills prior to 1955 as its method of solid waste disposal.

### Present Practices

St. Louis currently utilizes incineration as the major method of disposing of its solid wastes. The city operates two incinerators with each handling half of the city's production of 1000 TPD of domestic solid wastes. Private concerns are responsible for the city's commercial and industrial solid waste material. As a consequence of increasing air pollution, the U.S. Environmental Protection Agency has ordered the two incinerators closed by October, 1986.

### Future Practices

After the present incinerators are shut down, St. Louis will for approximately four years dispose of its solid wastes in landfills. During this time period, new incinerators will be constructed which will conform to EPA standards. These new incinerators or mass burning resource recovery facilities, will sell the generated steam to a utility company to be used for residential heating.

---

## A1-5 SOUTH ATLANTIC REGION

The South Atlantic region is made up of eight states and the District of Columbia. These states include: Delaware, Maryland, Virginia, West Virginia, North Carolina, South Carolina, Georgia, and Florida.

### Population Forecast

The populaton in the South Atlantic region is expected to increase from 28.5 million to 46 million in the 30-year period from 1965-1995. This marks a 61.4% increase. This is substantially above the projected increase for the nation for that same period.

## BALTIMORE (MD)

### Population Forecast

Baltimore has experienced a 13% decline in population over the last decade. In 1970, the population was 905,787. By 1980, it had dropped to 786,775.

### Solid Waste Generation and Management

Baltimore produces roughly 3000 tons of trash per day, or about 1.1 million tons per year. In the past, Baltimore has landfilled and incinerated its MSW. A pyrolysis project, the Monsanto Langard System, was attempted in the late 1970s, but failed mainly because of technological difficulties. Baltimore had the plant converted to a mass-burn facility by Signal Environmental Systems Incorporated. The mass-burn facility has been in operation since May of 1985. It has the capacity to incinerate 2250 TPD, but plans to upgrade its capacity are underway. About 75% of the MSW is incinerated at the mass-burn facility. It contains a waterwall boiler system which generates steam. The steam is either sold to the Baltimore Thermal Corporation, or is pumped to an on-site electricity turbine. The plant has a capacity to generate 60 megawatts of electricity, which is sold to the Baltimore Gas and Electricity Company. With an output of 420 million kilowatt-hours per year, the facility is able to supply the yearly power needs of more than 40,000 homes. Ferrous metals are recovered from the residual ash.

A private incinerator located in Baltimore City is owned and operated by the Polanski Company. It transports from Baltimore 300 TPD and has a 20-year contract.

Another 1000 TPD is incinerated without energy recovery. A very small amount of Baltimore's MSW (< 200 TPD) is landfilled. Approximately 800 to 1000 TPD of ash is produced by the incineration of the trash. All of the ash is landfilled.

### Future Plans

Baltimore will continue to incinerate most of its MSW for the coming years. A greater percentage of the MSW will be used for energy recovery when the capacity of the Signal Environmental System facility is increased. The city has a 20-year contract with Signal Environmental Systems Incorporated, and will decide on future management plans near the end of the contract period.

## ATLANTA (GA)

### Population Forecast

From 1970 to 1983 the population in Atlanta increased from 1,684,000 to 2,305,000. This total increase of 36.9%, represents an annual growth rate of approximately 2.8%. This growth rate is slightly above the 2% annual increase projected for the South Atlantic region in the 30-year period from 1965-1995.

### Solid Waste Generation

The volumes of solid wastes produced in Atlanta between 1969 and 1979 increased from 1,570 TPD to approximately 2,523 TPD. This marks an increase of 61%, or an annual increase of 6.1%.

### Past Practices

Atlanta relied on incineration and landfilling as the preferred methods of municipal solid waste disposal in the past. Wastes are collected by public agencies.

In the 1920s and 1930s the city of Atlanta relied heavily on incineration and recycling. These practices continued into the late 1960s and early 1970s. The incinerators were closed at that time because they could not meet the Environmental Protection Agency's air emission standards. The capital investment of tens of millions of dollars needed to upgrade the pollution control devices was not made; instead the incinerators were closed. This decision was made because the EPA would not give any guarantees on future emission standards. The city of Atlanta feared that after bringing the incinerators up to EPA standards, EPA would raise the standards. The potential financial risk was judged to be too great. Consequently, Atlanta moved from incineration to landfilling.

Availability of land close to the city makes landfilling a viable solid waste disposal option for Atlanta and additional land was recently purchased for new landfills.

### Present Practices

Inside the city there are four landfills which are owned by the city. The smallest of the four landfills has a remaining life span of five-to-seven years at the present rate of filling. The three larger landfills have remaining life spans of 10-to12-years.

Source separation centers, located at area shopping centers, have not proved to be an effective method of reducing the volume of waste, or the cost of managing them. The cost of source separation has actually increased the overall cost of solid waste disposal, while the dependency on local landfills remained almost the same.

**Future Plans**

Atlanta officials do not perceive the City's municipal solid wastes as a problem, but rather as an opportunistic fuel for generation of electricity in modern resource recovery facilities. Landfilling currently is the dominant method of solid waste disposal. This approach, coupled with minor amounts of source separation, represents the future solid waste disposal plan for Atlanta. Potential methods of solid waste disposal for the future include: (1) landfilling, (2) incineration, and (3) co-disposal of municipal solid waste and sewage sludge. Each has positive and negative impacts.

## MIAMI (FL)

**Population Forecast**

In the 13-year period from 1970-1983 the population of the Miami-Hialeah primary metropolitan area increased from 1,268,000 to 1,719,000. This increase marks a growth of 35.6% or an annual increase of 2.7%. This growth rate of 2.7% per year is slightly above the 2% annual growth projected for the South Atlantic Region during the 30-year period from 1965-1995.

**Solid Waste Generation**

The volumes of solid waste will increase from 5100 TPD to 7000 TPD in the 20-year period from the mid-1970's to the mid-1990's. This marks an increase of approximately 37.3%.

**Past Practices**

From 1929-1979, Miami relied heavily on sanitary landfilling. Eighty percent of the solid wastes produced in Dade County were disposed of in this way. The remaining 20% was incinerated. Six incinerators and two dozen landfills were in operation until 1970. After this time the numbers decreased markedly.

**Present Practices and Future Plans**

Increased environmental awareness in the late 1960s brought to light the potential threats posed by landfills and incinerators. Landfills in the area near the Biscayne aquifer posed an immediate threat to the county's drinking water. Incinerators presented problems with air pollution. Increased concern caused the closing of numerous landfills and incinerators, and brought anti-pollution controls to others. In 1979 serious planning was begun for solid waste facilities for the future.

Planning for systems to meet Dade County's serious waste disposal problem include two disposal facilities and three automated transfer stations. The transfer station on N.W. 20th Street (Northern Dade Transfer Station) handles the waste material for the city of Miami, while the other two transfer stations handle the wastes from the rest of Dade County. The new disposal facilities were chosen on the basis of environmental and economic concerns that took into consideration the increasing quantities and varieties of waste.

The Dade County Resources Recovery Plant (formerly called the Northwest Resource Recovery Facility of Dade County) was designed and located to service the northern portion of the county. The facility is located on a 160-acre site that was once the site of the main county landfill. Opening of the facility allowed for the use of the landfill to be phased out. In October 1982, when the facility reached design capacity, it was processing over 2500 tons of solid wastes per day. Electricity produced from this system supplied the needs for over 50,000 households in Dade County. The Northwest Resource Recovery Facility also recovers metals, aluminum, ferrous metals, and glass for sale to outside markets. Adjacent to the resource recovery facility there is a sanitary landfill for demolition debris, processed materials, and ash from the resource recovery facility. Only 3% by volume of the incoming waste is disposed of in the landfill. The leachate problem, an environmental concern often associated with landfills, is controlled by withdrawal of 1.5 to 1.8 million gallons of contaminated groundwater from the site. This water is later used in the resource recovery process. The garbage is mixed with water, pressed, dryed and pulped.

The Northwest Resource Recovery Facility was built through a contract with Parson's and Whittemore of New York through the Resource Recovery (Dade County) Subsidiary Construction Corporation. The facility was purchased by the county on 1 April 1983 after having met production and environmental test criteria. As of 20 June 1985 the county terminated contracts with the RRDC facility. Operation and maintenance of the plant was taken up by the Monteray Power Corporation on 21 June 1985.

The southern portion of Dade County is serviced by the South Dade Solid Waste Disposal Facility. This facility is equipped with a solid waste shredder, hazardous waste incinerator, sanitary landfill, and a maintenance operation. The capacity of the disposal site is 2100 TPD. The shredder, in operation since 1981, processes approximately 3000 TPD. The hazardous waste incinerator has a very small capacity of less than 100 TPD. Hospital wastes and pesticides are disposed of here. The landfill is used as a depository for shredded municipal solid waste, ash from the incineration process, and bulk wastes.

The sanitary landfill at this site is designed to prevent surface and groundwater pollution. It is located over a clay soil layer, has a supplemental underground drainage system, and is located over a salt intruded portion of the aquifer.

The remaining 100 TPD of solid wastes handled at this facility are incinerated in small-scale, experimental resource recovery plants.

In addition to two major disposal sites, three automated transfer stations were part of the 1971 solid waste management plan. According to the plan, these systems were operational by 1981. The three transfer stations are distributed throughout the city. Each is located on a phased out disposal site, or one that will be phased out as a result of the new solid waste management plan. The West Transfer Station, with a capacity of 1300 TPD, opened in March 1982 on a five-acre site adjacent to the old Coral Gables incinerator. It has been known to operate well above capacity, processing 2200 TPD. The Northeast Transfer Station is located at the site of the old Northeast Incinerator. The opening of the station allowed for the Northeast Incinerator, the North Landfill, and the Munisport Landfill to close. Located on nine acres of land, the station has a capacity of 1300 TPD. The capacity of 1300 TPD was achieved in March 1982 with the completion of Phase II. This station has processed up to 1900 TPD. The Central Transfer Station located in the Civic Center area opened in May 1978. This transfer station permitted the closure of the Virginia Key Landfill. This facility was designed to process 1200 TPD, but has been known to process a maximum of 2400 TPD.

Sites prescribed for the solid waste management plan for disposal facilities and transfer stations are at or adjacent to previously used sites. This ensured that there was no disruption of collection, no route changes, and no increased traffic.

Today, Dade County's resource recovery facilities are being taxed to their capacity with residents disposing an average of seven pounds per person per day. In the next several years, Dade County will be expanding their facilities to handle the additional load.

---

## A1-6 EAST SOUTH CENTRAL REGION

The East South Central region is made up of four states: Kentucky, Tennessee, Alabama, and Mississippi.

### Population Forecast

The East South Central region is expected to experience a population increase from 12.5 million to 16.5 million in the 30-year period from 1965-1995. This marks an increase of 32%. This is very close to the projected national increase for that same period.

## MEMPHIS (TN)

### Population Forecasts

In the 10-year period from 1970 to 1980 the population of Memphis has increased from 623,530 to 646,356. This is an increase in population of 3.7% or a 0.37% increase per year. This is less than the 1.1% projected growth rate for the East South Central region during the 30-year period from 1965-1995.

### Present Practices

Memphis relies on landfills for solid waste disposal. There are two landfills presently in operation. One is operated by the country (Shelby) and has an estimated life expectancy of two years. The other, owned by a private concern, has an estimated life expectancy of ten years. The county is in the process of obtaining an extension on the operation of their landfill while the private concern is in the process of obtaining permission to operate two additional landfills. An estimated 3000 TPD of solid waste material are collected, 40% by the city and 60% by private carters.

### Future Practices

Memphis is waiting to see whether the new landfills will be approved by the State. Plans are also being developed for the operation of a mass burning resource recovery facility with a 1500 TPD capacity, which would be operational in five-to-seven years.

## BIRMINGHAM (AL)

### Population Trends

In the 13-year period from 1970 to 1983 the population of the Birmingham metropolitan area has increased from 794,000 to 891,000. This is a increase in population of 12.2% or 0.94% increase per year. This growth rate is below the 1.1% projected growth rate for the East South Central region during the 30-year period from 1965-1995.

### Present Practices

Birmingham relies on landfills for solid waste disposal. Two landfills, with a life expectancy of 8- and 20-years respectively, accept about 1000 TPD of solid wastes.

### Future Practices

Presently, Birmingham does not recycle its solid waste material. There is currently a proposal before the City Council to construct a methane production facility, with a design capacity of two million cubic yards, located on a landfill. There are no reliable estimates of future solid waste volumes.

---

## A1-7 WEST SOUTH CENTRAL

The West South Central region is made up of four states: Arkansas, Louisiana, Oklahoma, and Texas.

### Population Forecast

Population of the West South Central region is projected to increase from 18 million to 30.5 million in the 30-year period from 1965-1995. This increase of 69.4% is substantially above the increase projected for the nation over this same period.

## Appendix 1

### DALLAS (TX)

#### Population Forecast

In the 13-year period from 1970-1983 the population of the Dallas-Fort Worth Metropolitan area increased from 2,352,000 to 3,266,000. This increase marks a growth of 38.9%, or an annual growth rate of approximately 3.0%. This growth rate is well above the 2.3% annual increase projected for the West South Central region in the 30-year period from 1965-1995.

#### Past Practices

Dallas produces about 1.5 million tons of wastes per year. The 2% per year increase typical of at least the past decade is expected to continue in the future. Dallas always has landfilled its municipal solid waste. There is no shortage of landfill space within economical distances of the city. A total area of 2,040 acres have been designated for disposal of the city's waste. The predicted lifespan is 75 years.

#### Future Practices

Dallas plans to continue landfilling. Since the life expectency of landfill space will extend well into the middle of the next century, there is no pressing need to implement resource recovery in their waste management program.

### NEW ORLEANS (LA)

#### Population Forecasts

In the 13-year period from 1970-1983 the population of New Orleans has increased from 1,100,000 to 1,361,000. This is an increase of 19.6% or 1.5% per year. This increase is below the 2.3% annual increase projected for the West South Central region in the 30-year period from 1965 to 1995.

#### Present Practices

New Orleans relies on landfills for its solid waste disposal. Approximately 1400 tons of solid waste is accepted each day at its two landfills (totaling 950 acres).

#### Future Practices

Landfilling will continue to be the primary method of solid waste disposal. But New Orleans has currently under study other solid waste disposal techniques such as mass burning resource recovery facilities.

---

## A1-8 MOUNTAIN REGION

The Mountain region is made up of five states: Montana, Idaho, Wyoming, Colorado, New Mexico.

### Population Forecast

The Mountain region is expected to experience a population growth of 133 percent over the 30-year period from 1965-1995. This increase is marked by an increase from 7.5 million to 17.5 million. This is substantially above the projected national increases for this same time period. The Mountain region is expected to have the largest population growth of any region in the Country during this 30-year period.

## DENVER (C0)

### Population Forecasts

In the 15-year period from 1970-1985 the population of Denver has decreased from 514,678 to 510,700. This is a decline in population of 0.7%, or a 0.05% decrease per year. For the years 1990, 2000, and 2010, however, Denver anticipates population increases of 1.8%, 6.4%, and 9.9% ,respectively, using 1970 as the base year.

### Present Practices

Denver relies on landfills for solid waste disposal. The city owns one of the landfills being used and three others are privately owned. The solid wastes are taken to a transfer station prior to disposal in landfills. The life expectancy of the privately owned landfills is about 20-years, while that of the city landfill is somewhat longer. About 250,000 tons of solid waste material is accepted each year at the landfills.

A "large item pick-up service" was established in 1985 which increased the amount of wastes collected. There appears to be a seasonal variation in the amount of waste, being 25% less in winter than in summer.

### Future Practices

No resource recovery facilities are planned in the foreseeable future. Based upon the past ten years of waste production, no large increase is anticipated in the volume of solid waste, but no in-depth studies have been undertaken to evaluate solid waste production. Denver has been unable to acquire additional land for the operation of more landfills for the past ten years because of the Downstone Amendment.

## ALBUQUERQUE (NM)

### Population Forecast

In the 14-year period from 1970-1984 the population of Albuquerque has increased from 243,751 to 350,475. This is an increase in population of 43.8%, or 3.1% increase per year. This is less than the 4.4% projected growth rate for the Mountain region during the 30-year period from 1965 to 1995.

**Present Practices**

Albuquerque relies on landfill for solid waste disposal. It leases a tract of land, from the State of New Mexico (100 acres, including a pit 30 feet in depth), as its landfill site. Approximately 1200 tons of solid waste is accepted each day at the site. The solid waste is compacted before burial. The city is trying to obtain an extension on its lease with the state, which expires in 1987.

**Future Practices**

Albuquerque and Bernalillo County, the county in which it is located, have formed a joint authority in creating a regional landfill. This new authority will also include the three or four adjacent counties. Requests for proposals have been issued to evaluate various resource recovery schemes and to evaluate the various types of equipment operated at the landfill site, such as shredding machinery, etc.

---

## A1-9 PACIFIC REGION

The Pacific region is made up of five states: California, Washington, Oregon, Alaska, and Hawaii.

**Population Forecast**

The population in the Pacific region is expected to increase from 24 million to 40 million in the 30-year period from 1965-1995. This marks a 66.7% increase. This is much greater than expected for the nation during the same time period.

### LOS ANGELES (CA)

**Population Forecast**

In the 13-year period from 1970-1983 the population of the primary metropolitan statistical area of Los Angeles-Long Beach increased from 7,042,000 to 7,818,000. This increase marks a growth of 11.0% or approximately 0.8% per year. This growth rate is substantially below the Pacific region's projected annual growth rate of 2.2% during the 30-year period from 1965-1995.

**Past and Present Practices**

Throughout the history of Los Angeles County, solid wastes have been disposed of in both publicly and privately owned sanitary landfills. Presently the county makes use of six collection yards: (1) South Central; (2) North Central; (3) Western; (4) East Valley; (5) West Valley; and (6) Harbor. In addition to the six collection stations, Los Angeles County operates four landfills: (1) Toyon Canyon, (2) Lopez Canyon, (3) Scholl Canyon, and (4) Calabscis Canyon; and one transfer station. The BKK transfer station is privately owned. Table A1-2 based on data from a report published by the Bureau of Public Works on 28 June 1983 lists fill rates.

Table A1-2 Solid Waste Disposal Fill Rates for Los Angeles County.
(From Bureau of Public Works, 28 June 1983).

| SITE | Operator | Fill Rate (TPD) | Other | Remaining Capacity (mil. cu yd.) | Closing Date |
|---|---|---|---|---|---|
| Toyon | City | 1650 | 1000 | 0.5 | 1984 |
| Lopez | City | 2300 | 600 | 11.1 | ** |
| Scholl | County | 200 | 2400 est. | 22.7 | 2011 * |
| Calabas. | County | 500 | 1000 est. | ---- | 2020 * |
| BKK TS | Priv. | 220 | ---- | ---- | 2003 * |

* Date of closure is based on the County's Solid Waste Management Plans' ultimate capacity estimates.

** The Ultimate Life of the Lopez Canyon landfill depends on the expansion of the Toyon Canyon landfill (Toyon II.)

The Toyon I landfill was closed at the end of 1985. Alternatives need to be formulated by 1993 or 1996 depending on the status of the Toyon II landfill. The city will be short 66% of the volume necessary for disposal in 1993 should the Toyon II landfill fail to open. Sixty-six percent of the volume represents the capacity of the Lopez Canyon landfill. If the Toyon II landfill is opened, the Lopez Canyon site will not reach capacity until 1996. The future of the Toyon II landfill is uncertain. It was denied conditional approval when the Mayor vetoed its operation.

**Future Plans**

According to the Long Term Landfill and Alternate Disposal Facility Policy, published on 28 June, 1983, Los Angeles County has to come up with long-term landfill disposal sites, and alternate disposal facilities.

A two phase study of solid waste disposal for Los Angeles was undertaken to determine the most economically and environmentally feasible means of disposal. Phase I of this study included

(1) discussion of the existing system,

(2) market analyses for steam and electricity,

(3) analysis of refuse composition and volume,

(4) technology appropriateness, and

(5) requirements for financing

Phase II, the more advanced planning stage, included studies of:

(1) technology determinations,

(2) potential sites,

(3) revenue streams and markets,

(4) financial programs and potential risk,

(5) environmental impact, and

(6) regulatory constraints

Use of private and publicly owned landfills is planned to be an interim measure. Landfills will continue to be used even after the opening of resource recovery facilities. They will be used as disposal sites for resource recovery ash, non-combustibles, and backup, should the resource recovery facility be down because of maintenance or breakdown. The city currently is trying to get the land necessary to expand the Lopez Canyon landfill by July 1987. This will not be allowed unless landfill alternatives are being planned.

The Lopez Canyon landfill is by far the most heavily relied upon solid waste disposal facility for the city. With current rates and methods, 66% of the refuse from the five areas will be disposed here by 1990. The expansion of the Toyon Canyon landfill will increase the life of the Lopez Canyon landfill by three years.

This study proposed three options for solid waste disposal in Los Angeles. The best plan is thought to be some mix of these options. These options include:

(l) reduction of waste volumes going to landfills through the implementation of material and energy recovery with modern resource recovery facilities.

(2) expansion and addition to the city owned and operated landfills; and

(3) expansion and addition of privately owned landfills.

The resource recovery option is two-fold. Materials recovery and energy recovery are both part of the overall plan.

## Materials Recovery

At the time of the study the pilot project for curbside collection had not begun. Implementation of this project, to service 15,000 homes in West Los Angeles, was scheduled for fall 1983. This program calls for biweekly curbside collection of metals, glass, and newspapers. These materials represent 19% by volume of the refuse collected. According to projections, because of the voluntary nature of the project, only 25% of this 19% will be recovered. This will lead to an overall reduction in waste volume of 4%-5%. Part of the materials recovery program includes design of a dual purpose garbage trucks. This would allow for simultaneous refuse and recyclable collection.

## Energy Recovery

A two phase feasibility study was conducted by the Bureau of Sanitation for a Central City Waste-to-Energy System. This system will be designed to handle 1600 TPD. The questions of each phase were presented earlier. Results of the studies favor several 800 TPD mass burn facilities. These facilities would ease the load on the Toyon Canyon landfill. When these facilities reaches capacity wastes will have to be hauled to the Lopez Canyon landfill. Resource recovery with its ability to greatly reduce volumes will ease the load on the Toyon Canyon facility, increase the life expectancy of this site and the Lopez Canyon site, and cut down on haul distances.

The City has recently (1986) signed a contract with Ogden-Martin Systems for the construction of a 1600 ton per day resource recovery facility.

## SEATTLE (WA)

### Population Forecast

The population of the consolidated metropolitan statistical area of Seattle-Tacoma increased from 1,837,000 to 2,187,000 in the 13-year period from 1970-1983. This increase marks a population growth of 19%, or an annual increase of 1.5%. This increase is below the Pacific region's projected annual growth rate of 2.2% during the 30-year period from 1965-1995.

### Solid Waste Generation Forecast

The volumes of solid waste produced in Seattle are expected to grow as a result of the increased employment in the city. Volumes are expected to increase from 368,000 TPY in 1985 to 550,000 TPY in 2000.

## Appendix 1

### Past Practices

Seattle is concerned with decreasing the volumes of solid wastes. This concern stems from the limited availability of land for disposal of solid wastes within the city. The city of Seattle is serviced by two landfills: (1) Midway; and (2) Kent Highlands. The Midway landfill was closed in October, 1983. Kent Highlands will be closed by the end of 1986.

Incentive programs for waste reduction were implemented in the form of variable beverage can rates in 1981. Nineteen percent of the wastes produced in Seattle in 1982 were recycled. This figure is expected to rise over the years. Plans for solid waste reduction include volume reduction and resource recovery.

### Present Practices and Future Plans

New and additional methods of solid waste disposal are being studied in Seattle. The options being researched include:

(1) waste reduction,

(2) landfilling,

(3) material recovery, and

(4) energy recovery.

The King County Cedar Hills landfill located 20 miles southeast of Seattle is much farther from the city than the other two landfills that Seattle uses or has used. Haul distances of this length will increase transportation costs by 66.6%.

Waste reduction and resource recovery are part of a program to decrease dependence on landfilling. By 1986, the fraction of solid waste recycled is expected to increase from 19% to 40%. Other methods of recycling that are being used include

(1) composting,

(2) buy-back centers, and

(3) drop-off centers.

Recycling is probably encouraged and practiced in Seattle more than in any other United States city.

# APPENDIX 2

## Extant Regulatory Considerations Regarding Air Quality

### A2-1 AIR QUALITY CRITERIA

In response to growing public concern over the quality of the nation's air, Congress passed the Clean Air Act in 1970. In order to have some measure of what constitutes clean air, it was necessary to establish a set of criteria representing the maximum level of certain pollutants. These pollutants, defined as criteria pollutants, include sulfur oxides, carbon monoxide, nitrogen dioxide, ozone, lead and particulate matter. The allowable level of each of these is specified in the National Ambient Air Quality Standards (NAAQS) summarized in Table A2-1.

Other pollutants than those specified in the NAAQS standards are also regulated at the federal level under the Clean Air Act. Since these pollutants are not taken into account in defining what is clean air, they are called non-criteria pollutants. Table A2-2 lists pollutants in that class.

In order to monitor compliance with the provisions of the Clean Air Act the United States is divided into 247 air quality control regions (AQCRs). There are provisions for an in depth review of any new proposed industrial activites that might generate significant pollution, and to review existing sources. A source is considered to be significant if it could emit over 250 tons per year of any of the regulated pollutants specified in Table A2-1 or if it is one of a number of prescribed industrial activies (known to be major sources of air pollution) and has the potential of emitting more than 100 tons of regulated pollutants per year. Sources failing to meet these requirements must be subjected to a Prevention of Significant Deterioration (PSD) regulation review.

Any source whose emissions fall within the acceptable zones set by the PSD regulations but exceed the specific emission limits of one of the regulated pollutants in Table A2-1 must be subjected to a review under the Best Available Control Technology (BACT) regulations.

The review process for a the creation of a new facility depends upon several factors. One has to do with whether the facility is to be sited in a geographical region that is already in compliance with the NAAQS and whether the facility qualifies as a significant new source. If it is not a significant new source it does not fall under the federal PSD regulations and must only satisfy any applicable state regulations, which in some cases may be more restrictive than the federal NAAQS standards.

If the criteria of the Clean Air Act lead the proposed new facility to be judged as a major new or modified facility then the next question to be asked is whether the facility will be located in a geographical area in compliance with NAAQS or not. If the area is already in compliance, the new facility must be reviewed to determine if the emissions are below half of those allowed by the NAAQS. If so, the permit request can go directly to the public hearing phase. If the emissions are projected to exceed the 50 % NAAQS level, a much more complex analysis must be made to determine what emission reduction steps can be instituted to insure that the

NAAQS guidelines will be met. If this can be shown, then the permit request may then proceed to the public hearing phase.

If the "significant" new or modified facility is to be located in an area that is not in compliance with the NAAQS a new policy comes into play, that of the Emissions Offset Policy (EOP). This policy is intended to insure that every effort is made to control emissions to the greatest possible extent, that more than offsetting emission reductions are achieved from exisiting sources within this same physical site, and that adequate progress will be made toward the full achievement of the NAAQS standards. The review of plans developed under the EOP must be examined and approved by the EPA and the appropriate state agency prior to the calling of a public hearing on the permit request.

## A2-2 POLLUTION STANDARDS INDEX

It is difficult to get an overall assessment of the quality of air in selected locations from an examination of the NAAQS criteria. One location may have essentially no oxides or particulate matter but have large concentrations of ozone, while another area may have the reverse situation. Yet, the true quality of the air in these two locations, from the point of view of human health, may be quite different. In order to attempt to quantify the overall quality of air the Pollutant Standard Index (PSI) has been established. It is based on the weighted concentrations of the criteria pollutants specified in the NAAQS.

The range of values possible for the PSI is 0 to 500. The value zero is the absolute optimum, while 500 is the absolute worst possible quality. Any rating above 300 is considered hazardous. Scores between 200 and 299 are regarded as "very unhealthful", between 100 and 199 as "unhealthful", between 50 and 99 as "moderate", and below 50 as "good". Table A2-3 reveals how various metropolitan areas of the country rank relative to the PSI index.

As shown in Table A2-4 there is some variability among the various polluting sources regarding the extent to which they comply with the NAAQS standards. Only 46 % of the primary smelters are in compliance, whereas 97 % of the coal cleaning plants are in compliance. Evidence indicates that in most metropolitan areas exhaust from automobiles is the primary factor in determining the overall air PSI index and that emissions from incinerators make only a small contribution.

**Table A2-1 National Ambient Air Quality Standards for the criteria pollutants.**

| Criteria Pollutant | Averaging Time | Primary Standard | Secondary Standard |
|---|---|---|---|
| | | (in units of micrograms per cubic meter) | |
| Particulate Matter | Annual | 75 | 60 |
| | 24 hr* | 260 | 150 |
| Sulfur Oxides | Annual | 80 | |
| | 24 hr* | 365 | |
| | 3 hr* | | 1,300 |
| Carbon Monoxide | 8 hr* | 10,000 | 10,000 |
| | 1 hr* | 40,000 | 40,000 |
| Nitrogen Dioxide | Annual | 100 | 100 |
| Ozone | 1 hr* | 240 | 240 |
| Hydrocarbons (non-methane and 6 AM- 9AM) | 3 hr | 160 | 160 |
| Lead | 3 months | 1.5 | 1.5 |

* These levels are not to be exceeded more than once each year.

**Table A2-2 Noncriteria pollutants**

| | |
|---|---|
| Asbestos | Total Reduced Sulfur |
| Beryllium | Reduced Sulfur Compounds |
| Mercury | Hydrogen Sulfide |
| Fluorides | Sulfuric Acid Mist |
| Vinyl Chloride | |

Source: "Prevention of Significant Deterioration" In *US EPA Workshop Manual* (Washington, DC: Government Printing Office, October 1980).

**Table A2-3 Ranking of Standard Metropolitan Areas according to the Pollution Standards Index (1976 - 1980) [Ref. 14 in Ch. 4].**

| Standard Metropolitan Statistical Area | Number of Days Per Year with index above 100 | with index above 200 |
|---|---|---|
| Los Angeles | 242 | 118 |
| New York | 224 | 51 |
| Pittsburgh | 168 | 31 |
| San Bernardino | 167 | 88 |
| Clevland | 145 | 35 |
| St. Louis | 136 | 29 |
| Chicago | 124 | 21 |
| Louisville | 19 | 12 |
| Washington, DC | 97 | 8 |
| Phoenix | 84 | 10 |
| Philadelphia | 82 | 9 |
| Seattle | 82 | 4 |
| Salt Lake City | 81 | 18 |
| Birmingham | 75 | 19 |
| Portland | 75 | 3 |
| Houston | 69 | 16 |
| Detroit | 65 | 4 |
| Jersey City | 65 | 4 |
| Baltimore | 60 | 12 |
| San Diego | 52 | 6 |
| Cincinatti | 45 | 2 |
| Dayton | 45 | 2 |
| East Chicago | 36 | 8 |
| Indianapolis | 36 | 2 |
| Milwaukee | 33 | 6 |
| Buffalo | 31 | 5 |
| San Francisco | 30 | 1 |
| Kansas City | 29 | 6 |
| Memphis | 28 | 2 |
| Sacramento | 28 | 2 |
| Allentown | 27 | 1 |

**Table A2-4 Compliance status of major air pollution sources (1980) [Ref. 14 in Chapter 4].**

| Industry | Total Number of Sources | Number In Compliance | Percent In Compliance |
|---|---|---|---|
| Coal Cleaning | 409 | 395 | 97 |
| Asphalt Concrete | 2862 | 2752 | 96 |
| Sulfuric Acid | 262 | 246 | 94 |
| Phosphatic Fertilizers | 69 | 62 | 90 |
| Portland Cement | 200 | 176 | 88 |
| Gray Iron | 433 | 381 | 88 |
| Pulp and Paper | 475 | 417 | 87 |
| Municipal Incinerators | 72 | 60 | 83 |
| Power Plants | 700 | 559 | 80 |
| Petroleum Refineries | 214 | 170 | 79 |
| Aluminum Reduction | 49 | 37 | 76 |
| Iron and Steel | 204 | 110 | 54 |
| Primary Smelters | 28 | 13 | 46 |

# APPENDIX 3

## MUNICIPAL SOLID WASTE POLICY FORUM

**Results and Conclusions**

**of a Forum**

**J.R. Schubel and H.A. Neal**

**Conveners**

**24 January 1986**

**Report of**

**The Waste Management Institute**

**Marine Sciences Research Center**

**State University of New York at Stony Brook**

Appendix 3

## INTRODUCTION

The Municipal Solid Waste (MSW) Policy Forum held on 24 January 1986 was the second in a series of such fora sponsored by Stony Brook's Waste Management Institute. The Agenda for the Forum is contained in Appendix A; the list of participants in Appendix B.

These symposia were designed to bring together small groups of knowledgeable people to explore a wide range of municipal solid waste management issues. This particular Forum concentrated on the residuals--emissions and ash--from mass burn resource recovery facilities and was designed to give several of the major resource recovery industries an opportunity to present their assessment of state-of-the-art technology and the characteristics and levels of the residuals--emissions and ash--that can be achieved with modern plant design and proper plant operation. The Forum also provided an opportunity to identify research needs and opportunities, and to discuss alternative approaches to conducting this research. The New York State Energy Research and Development Authority's municipal solid waste research program was described in detail ( Appendix C, available on request).

This report summarizes those major findings and recommendations which emerged from the discussion which are particularly pertinent to Long Island and the Metropolitan New York City area. While all participants had the opportunity to review and comment on this document before printing, it does not necessarily follow that all participants endorse all of the findings and recommendations presented here. There was broad concensus, however, on all statements.

## FINDINGS AND CONCLUSIONS

### GENERAL

- o The per capita production of MSWs is higher in the United States than in any other country in the world; averaging nearly 5 pounds per person per day. On Long Island, the figure is nearly 6 pounds per person per day.
- o Every day New York State produces more than 40,000 tons of MSW.
- o The relative contributions of different kinds of wastes to the MSW stream are summarized in Figure A3-1.
- o Municipal solid waste--garbage and trash--presents a risk; MSW is itself a potential pollutant.
- o This garbage and trash must be disposed of.
- o The alternatives available for disposal of garbage and trash are

limited in number and in variety. Each has advantages and disadvantages. None is ideal; not even recycling. All entail risks.

- o The best--most appropriate--disposal strategy is the alternative which minimizes risk to public health and to the environment at acceptable cost, both in the short term and in the long term.

- o Active source reduction and recycling programs could reduce the volume of MSW requiring disposal, but not eliminate it. Such programs also could change the character of the ultimate waste product to make it more innocuous. In addition, source reduction and recycling programs conserve valuable natural resources, reduce pollution, and save energy.

- o Municipalities should develop comprehensive waste management strategies. Source reduction and recycling are appropriate and desirable components of such strategies.

## INCINERATION

- o Burning has inherent advantages as a method for garbage and trash disposal because of its purification properties, and because it reduces the amount of residual waste.

- o Municipal solid waste is not the best fuel; neither is it the worst.

- o If all of New York's MSW were burned in resource recovery facilities, it would generate 500 megawatts of electricity.

- o The selective removal of certain components from MSW before combustion may reduce risk to human health and the environment. Removal of batteries, for example, could reduce levels of nickel (Ni), cadmium (Cd), mercury (Hg) and lead (Pb).

- o Combustion technology used in modern resource recovery facilities constructed by major vendors represents a significant evolution from earlier designs typical of older incinerators.

- o Municipal solid waste burns readily but possesses a number of negative characteristics including: (1) heterogeneity in composition and particle size, (2) relatively low heating value (3800-5000 BTU/lb.), (3) relatively high chlorine ($Cl_2$) content (0.5%), (4) low ash fusion temperature, and (5) high ash and moisture content of fuel.

- o The basic principles of good combustion are described by the three "Ts"--Time, Temperature, and Turbulence. Time: the longer a particle is held at a high temperature, the more complete the combustion. Temperature: the higher the temperature, the more complete the combustion. Turbulence: the better the mixing, the greater the likelihood of getting oxygen ($O_2$) (air) to each waste element thus enhancing the completeness of combustion.

- There are several diagnostic indicators of good combustion:

  (1) low emissions of carbon monoxide (CO), hydrocarbons and oxides of nitrogen ($N_2$);

  (2) very low content of carbon (C) and combustible material in the ash residue; and

  (3) boiler efficiency.

- Two primary goals of incineration are to maximize combustion and minimize air pollution. There are two other goals: high plant availability (absence of shutdowns) and low facility maintenance cost.

- The principal indicators of incomplete combustion are high levels of $CO_2$ and CO. Carbon monoxide is an air pollutant and contributes to the corrosion of boiler surfaces. Carbon monoxide is an indicator of the presence of other products of incomplete combustion.

- Conditions for and characteristics of good combustion in resource recovery facilities include:

  (1) a hot uniform firebed devoid of cool and hot spots on the grate;

  (2) an adequate secondary air supply mixed thoroughly into the hot fire gases rising from the fire bed;

  (3) flue gas temperatures at, or above, $1600^{o}F$ for approximately 1 second after the flue gas leaves the secondary firing zone; and

  (4) avoiding combustion upsets on the grate or in the second firing zone.

- Evidence that good combustion has been achieved and maintained is manifested in the flue gas by a steady 7-10% oxygen level and less than 100 ppm CO.

- Carbon monoxide and oxygen can be monitored continuously, although CO monitoring is difficult.

- Continuous monitoring of CO is the best single method for assessing how well a plant is operating.

- Products of incomplete combustion include a wide range of organics and particulates.

- As combustion becomes more complete, $Cl_2$ produced from burning of organochlorines is converted to HCl. This is the desired fate for $Cl_2$ since it can be removed with scrubbers.

- More than 700 compounds have been identified in the emissions of resource recovery facilities.

- The U.S. Environmental Protection Agency and many state environmental and health agencies are developing criteria to assess the quality of combustion in resource recovery facilities.

## ASH

- The incineration of garbage produces large amounts of ash which must be disposed of. The problem of how to accomplish this disposal will be particularly acute on Long Island and in the metropolitan New York City area. More than eleven million tons of solid wastes are collected annually. This translates into a potential of more than 2.2 million tons of incinerator ash each year; enough ash to make more than 65 million cinder block-size blocks each year.

- Fly ash accounts for about 5-10% of the total ash residue from a modern resource recovery facility; the remaining 90-95% is bottom ash.

- The relative contributions of different kinds of wastes to the total MSW ash stream are summarized in Fig. A3-1.

- Most of the fly ash produced is removed from the stacks with electrostatic precipitators or baghouse filters. The particles are very fine, ranging from less than 1 micron (0.00004 in.) in diameter to about 500 micron (0.02 in.).

- Bottom ash drops through the grate where it is collected. Most particles range from about 0.04 in. to 0.4 in. in diameter. In addition, there may be larger pieces ranging from bottles and cans to automobile engines.

- Cadmium, lead, and several other metals vaporize during combustion and most precipitate out onto particulates in the stacks.

- Cadmium and lead can not be segregated effectively from municipal solid wastes because of the variety of wastes products in which they are found.

- o Leaching of landfilled, unstabilized ash from resource recovery plants is a function of a variety of physical and chemical properties including: permeability and porosity of the ash deposit, the frequency of deposition of ash and pH of the precipitation and interstitial waters.

- o The mixing of ash and MSW in a landfill promotes leaching of a number of metals, particularly Pb and Cd, from the ash. The decaying organic matter reduces the pH of pore waters and, as a result, accelerates leaching when the buffering capacity of the ash is exhausted.

- o The leaching rates of Cd and Pb from MSW ash increase with decreasing pH of precipitation and pore waters.

- o While only relatively small fractions of the Cd and Pb in unstabilized fly ash are available to the environment, those fractions which are, may be leached rapidly.

- o The elemental concentrations of metals in ashes--fly and bottom--of MSW are presented in Table A3-1. Concentrations of the same metals in coal ash are shown for comparison. Note the significant enrichment in Cd and Pb in MSW fly ash relative to coal fly ash. The total metals concentrations in either kind of fly ash are not available to the environment through leaching.

- o Scientists at Stony Brook's Marine Sciences Research Center have successively stabilized a variety of mixtures of fly ash and bottom ash from resource recovery facilities with Portland cement (~15%) into blocks which meet ASTM standards for construction.

- o Stabilization of fly ash can reduce markedly the potential for leaching of contaminants.

## DIOXINS AND FURANS

- o As the threshold of our ability to measure dioxins and furans has progressively gone down to lower and lower concentrations, these compounds have been found with increasing frequency.

- o It now is possible to detect dioxins in the parts per trillion range. To visualize a concentration in the part per billion range consider that looking for a single individual among the world's population today would be looking for 1 in 4.5 billion. A concentration of one in a trillion would be equivalent to picking out a single second in the last 32,000 years.

- o Dioxins and furans recently were found in Milorganite sealed in glass vials in 1933 and exhibited at the 1939 New York World's Fair. They were detected recently in sediments in Lake Huron which have been dated at 80 years old.

- o These observations and many others indicate that dioxins and furans have existed in the environment for a long time.

- o Data also indicate that the environmental levels of dioxins and furans increased significantly after chlorinated hydrocarbons became important industrial chemicals.

**Table A3-1 Elemental concentrations of metals in ashes from the incineration of municipal solid waste (MSW) and coal.***

| | MSW ASH(mg/kg) | | COAL ASH (mg/kg) | |
|---|---|---|---|---|
| ELEMENT | FLY ASH | BOTTOM ASH | FLY ASH | BOTTOM ASH |
| Ca | 54,500 | 50,500 | 45,000 | NR |
| Sr | 200 | 250 | 775 | 800 |
| Ba | 800 | 800 | 991 | 1,600 |
| Cd | 470 | <100 | 2 | 1 |
| $SiO_2$ | 319,000 | 368,000 | 483,000 | NR |
| Al | 70,000 | 33,000 | 92,000 | NR |
| Fe | 17,500 | 132,000 | 35,000 | NR |
| Ti | 14,600 | 3,600 | 19,400 | NR |
| Pb | 5,200 | 900 | 67 | 7 |
| Cr | 400 | 500 | 136 | 120 |

NR = Not Reported

*Courtesy of Signal Environmental Systems

- Existing laboratory data for rats, mice and several other small mammals indicate that 2,3,7,8 tetrachlorodibenzo-p-dioxin (TCDD) is one of the most acutely toxic anthropogenic materials known.

- To date over 40 municipal solid waste burning plants in at least 9 countries have been tested for dioxins and furans in bottom ash, in fly ash and in flue gas. Dioxins and furans have been found in all plants tested except one. The exception is a facility in Ames, Iowa. This facility burns a mixture of about 10% (by weight) RDF (refuse-derived fuel) and 90% pulverized coal at a temperature hotter than is conventional in modern resource recovery facilities.

- The levels of dioxins and furans emitted from municipal solid waste incinerators varies widely among the plants tested (Table A3-2).

- The levels of emission of dioxins and furans from mass burning of garbage and trash can vary from plant to plant by a factor of more than 1000 depending upon plant design, construction, and operation (Table A3-2).

- The differences in emissions shown in Table A3-2 can be attributed to a variety of factors. Some plants are old; others new. Some have furnaces with refractory walls; others have water-cooled walls. Some were field erected; others were not. Some are small; others are large. Some recover heat; others do not.

- The data in Table A3-2 indicate that facilities which recover heat tend to have lower emissions of dioxins and furans than those that do not. One exception is the Hamilton (Ontario) plant. This plant is of an old design and had been poorly maintained. A second exception is the Hampton (Virginia) plant which also is poorly designed and was poorly operated.

- The aggregation of emission data from incinerators and resource recovery facilities without discriminating between old and new plants, between well-designed and poorly-designed plants, and between well-operated and poorly-operated plants produces misleading results.

- In analyzing emission data, it is appropriate, and indeed desirable, to separate plants by age and design.

- Data such as those in Table A3-2 contain important information which can be useful in making scientific judgements about the levels of dioxins and furans achievable in modern resource recovery facilities, and in making management decisions regarding such facilities. That information is lost, however, if the data are simply averaged without distinguishing among differences in design and operation of the facilities from which the data were collected. Averaging emissions from well-tuned 1986 automobiles equipped with emission control devices along with emissions from Model T's and poorly maintained 1949 Studebakers will not provide an accurate estimate of emission levels achievable with modern automobile technology.

**Table A3-2 Dioxin (PCDD) stack emission data.***

| FACILITY (Country) | EMISSION RATE ($ng/m^3$) ALL PLANTS | HEAT RECOVERY PLANTS |
|---|---|---|
| STAPELFELD (Germany) | 31 | 31 |
| CHICAGO N.W.(USA) | 42 | 42 |
| ESKJO(Sweden) | 73 | 73 |
| STELLINGERMOOR(Germany) | 101 | 101 |
| PEI (Canada) | 107 | 107 |
| ZURICH (Switzerland) | 113 | 113 |
| BORSIGSTRASSE(Germany) | 128 | 128 |
| COMO (Italy) | 280 | 280 |
| ALBANY (USA) | 316 | 316 |
| DANISHRDF(Denmark) | 316 | 316 |
| ITALY 1 | 475 | - |
| ITALY 6 | 569 | - |
| BELGIUM | 680 | 680 |
| ITALY 5 | 1020 | - |
| ZAANSTAD (Holland) | 1294 | - |
| VALMADRERA(Italy) | 1568 | 1568 |
| HAMILTON(Canada) | 3680 | 3680 |
| HAMPTON (USA) | 4250 | 4250 |
| ITALY 4 | 4339 | - |
| TORONTO (Canada) | 5086 | - |
| ITALY 3 | 7491 | - |
| ITALY 2 | 48,808 | - |

*Source: Kay Jones, Roy F. Weston, Inc., Courtesy BFI, Inc. Plants are arranged in increasing order of emission of PCDD

- The data in Table A3-2, and other data, demonstrate that emissions of dioxins from the stacks of modern, well-designed and well-operated resource recovery facilities are likely to be below 150 nanograms per cubic meter ($ng/m^3$) of effluent.

- Good combustion minimizes the generation of dioxins and furans in modern resource recovery facilities.

- Effective removal of particulates from the flue gas further reduces the release to the air of dioxins, furans and other organic compounds and metals, especially if the stack temperature is low.

- Effective scrubbing of the flue gas reduces emissions of acid gases to the air.

- Application of existing technology can reduce emission of particulates and acid gases from modern resource recovery facilities to mandated levels.

- With good combustion in a modern resource recovery facility the emissions of dioxins and furans from the stack per ton of MSW incinerated may still be about 10X the amount on the fly ash recovered by the air pollution control system and 100X the amount contained in the bottom ash.

<u>Resource Recovery Facilities and Existing Guidelines and Standards</u>

- The U.S. Environmental Protection Agency (EPA) does not have official guidelines or standards for dioxin and furan emissions from resource recovery facilities, but is in the process of developing emissions criteria for these compounds.

- Guidelines for dioxin and furan emissions have been issued by Ontario, the Netherlands, Sweden and Denmark.

- Other countries, New York and other states in the U.S. are considering issuing guidelines for emissions of dioxins and furans from resource recovery facilities.

- At present, EPA and New York State comply, at least unofficially, to the guidelines set forth in the EPA's 1981 Hernandez document.

- The most stringent guidelines are those set forth in the Hernandez document and adopted by the EPA and New York State. If these guidelines have not been officially adopted by New York and EPA, they can at least be considered to be foster children. These guidelines are about 20X more stringent than Ontario's and 1000X more stringent than those of the Netherlands.

- o According to the data in Table A3-2, the first seven or eight plants would meet the New York State and EPA "guidelines" and several more would meet the Ontario guideline.

Dioxins and Furans--Summary

- o Most effective control of emissions of dioxins and furans from resource recovery facilities can be achieved through a combination of good combustion and effective removal of particulates from the flue gas. Scrubbing at low temperature has been shown to be particularly effective.

- o Application of state-of-the-art combustion technology in modern resource recovery facilities can reduce emissions of dioxins and furans to levels below the most stringent guidelines now in effect.

- o Routine monitoring techniques do not now exist for direct, continuous measurements of dioxins, furans, and other organic compounds in the flue gas.

- o Techniques do exist, however, to monitor the effectiveness of combustion.

- o The available data indicate that properly designed and operated resource recovery facilities can meet the emissions criteria used by New York and the EPA for dioxins and furans.

- o Most emission data for resource recovery facilities represent snapshots of instantaneous to short-term (a few hours) conditions taken at infrequent intervals. More data are needed to establish the variability of emissions among facilities and to establish the temporal variability of emissions at individual facilities over a range of seasons and operating conditions.

- o More data are needed to demonstrate that resource recovery facilities meet these criteria on a continuing basis.

- o There are in the world today several hundred large-scale and thousands of small-scale (apartment house) municipal solid waste incinerators that do not meet modern design and operating specifications.

- o Burning garbage and trash to produce energy is a good idea if the combustion is done in a modern, well-designed, well-maintained and well-operated facility.

- o To achieve the lowest emission levels, resource recovery facilities not only must be properly designed, but must also be properly maintained and operated.
- o The effective operation of sophisticated modern resource recovery facilities should be in the hands of well-trained operators.
- o Proper combustion can significantly reduce the emission levels of most contaminants of concern from resource recovery facilities.
- o Proper plant design does not guarantee that the plant will operate at or near design criteria.
- o Training for resource recovery facility operators should be mandated by the New York State Department of Environmental Conservation.

<u>Some Ways to Improve Management of Municipal Solid Waste</u>

- o A comprehensive municipal solid waste management program which incorporates resource recovery is not incompatible with source reduction and recycling. Indeed, the strategies can be complementary.
- o Construction of a modern, well-designed facility that is poorly maintained and operated does not represent an achievement for technology or society.
- o Disposal of hospital wastes may pose a greater public health threat than garbage and trash because of microbiological contamination.
- o Operation of resource recovery facilities by the private sector may have advantages over operation by the public sector. If the enforcer is not the operator, appropriate enforcement is more likely.
- o Contracts for operation of resource recovery facilities can be written to require the operator to handle the municipality's garbage and trash in the event of shutdowns--planned or unplanned.
- o Permitting can and should be used to ensure that resource recovery plants operate within the design envelope and, as a result, keep emissions within an acceptable range.
- o The permitting process is sufficiently flexible that many societal concerns can be accommodated and alleviated in the permitting and licensing procedures.
- o Arrangements should be made to accommodate a municipality's gar-

bage and trash during short periods when its resource recovery facility is shut down for planned or unplanned reasons. Options include landfilling and transfer to other resource recovery facility.

- Failure to make rigorous comparative assessments of the environmental and public health effects of the different disposal alternatives has been a major drawback in selecting the best--most desirable--strategy.

- Proper environmental assessments of different disposal strategies must include cross media (air-land-water) assessments. To date they have not.

- The configuration of existing Federal agencies makes rigorous and well balanced cross media analysis exceedingly difficult and improbable. Agencies are aligned along lines of each individual medium creating competition among units to protect turf, rather than to select the most desirable alternative. A total ecosystem approach is needed.

- At the present time, integration of municipal solid waste management programs at the federal level is weak and ineffective.

- Major changes in the permitting process are needed to ensure selection of the best alternative to manage municipal solid waste. Multimedia assessments are required.

- This situation could be resolved with an organic environmental law which focuses attention on the total ecosystem and requires cross-media analysis.

## RESEARCH NEEDS

### General

- A critical assessment is needed of the impacts of sanitary landfills on the total environment. Existing assessments have neglected the effects of landfills on the air. Information is needed both on gases and on particulates and adsorbed contaminants. This information is needed to compare and contrast the landfilling option with the resource recovery option.

- Research is needed to evaluate the environmental effects of land disposal of resource recovery ash (fly and bottom) and flue gas scrubber products and to develop techniques to mitigate any undesirable effects.

- Research is needed to assess the environmental and public health effects of disposal in the ocean of stabilized and unstabilized ash from resource recovery facilities. Questions concerning the products of leaching and their ecological effects, and the stability of both the ash and leachates should be studied under different environmental conditions in the laboratory and in the field.

- Additional research is needed to resolve uncertainty as to the locations and strengths of other sources of dioxins and furans to the environment, and to improve our understanding of the fates and effects of these families of compounds in the environment and on public health.

Resource Recovery

- A critical assessment is needed to determine which type of particulate control device--electrostatic precipitators or baghouse filters--are most effective in controlling particles and particle-bound contaminants.

- Additional research is needed to define the time-temperature conditions to promote adsorption of contaminants onto particles.

- Additional research is needed to obtain real-time measurements of the combustion process using physical sensing techniques such as fourier-transform infrared spectroscopy and Raman spectroscopy.

- Additional research is needed to test the efficacy of $O_2$-enrichment as a method of enhancing completeness of combustion.

- An accelerated research and development effort is needed to develop creative uses of ash from resource recovery facilities; uses which are safe and beneficial to society.

## A FINAL MESSAGE

Most participants agreed that further forums on the subject of municipal solid wastes would be useful and urged that the themes suggested for future forums included: risk assessment of different municipal solid waste management strategies; information on resource recovery for decision makers; source reduction and recycling; and examination of municipal solid waste management alternatives; and reconciling the differences between real and perceived public health risks of dioxins and furans from modern resource recovery facilities.

---

## SYMPOSIUM REPORT APPENDICES

A. Agenda

B. List of participants

C. The New York State Energy Research and Development Authority's Resource Recovery Research Program (available upon request)

Appendix 3

# MUNICIPAL SOLID WASTE POLICY FORUM

24 January 1986

Challenger Hall 165
Waste Management Institute
Marine Sciences Research Center
State University of New York at Stony Brook

| | |
|---|---|
| 0930 | Welcome and Introductions (Homer A. Neal, J.R. Schubel) |
| 0945 | An Overview of What We Hope to Achieve Today (J.R. Schubel) |
| 1000 | Environmental Concerns and Emissions from Resource Recovery Facilities (D. Sussman, Ogden Corp.) |
| 1030 | Dioxins (Clinton Kemp, BFI) |
| 1100 | Designing for Good Combustion (A. Licata, Dravo Energy Resources) |
| 1200 | Lunch |
| 1230 | MSRC's Ash Research Program (F. Roethel, MSRC) |
| 1245 | Emissions and Ash from Modern Mass Burn Resource Recovery Facilities: An Overview of Unresolved problems and Unexploited Opportunities (G. Smith, EPA) |
| 1300 | Management of Residues from Resource Recovery (M.R. Surgi, Allied Signal, Inc.) |

Solid Waste Management and the Environment

1330 An Overview of the N.Y. State Energy Research and Development Authority's (NYSERDA) Resource Recovery Research Program (Parker Mathusa, NYSERDA)

1400 Discussion and Formulation of Conclusions and Recommendations

## Appendix B

## LIST OF PARTICIPANTS

1. Ann Anderson, Senior Engineering Technician, New York State Department of Conservation, Region 1.
2. Harold Berger, Director, Region 1, New York State Department of Environmental Conservation.
3. Marc David Block, Co-Director, Science and Decision Making Project, New York Academy of Sciences.
4. Gerald Brezner, Regional Solid and Hazardous Waste Engineer, New York State Department of Environmental Conservation, Region 1.
5. Maggie Clarke, Environmental Scientist, New York City Department of Sanitation.
6. Terrence Curran, Executive Director, New York State Environmental Facilities Corp.
7. Norman G. Einspruch, Dean, College of Engineering, University of Miami.
8. Robert J. Fitzpatrick, Vice President, Grumman Corp.
9. Ted Goldfarb, Associate Professor of Chemistry and Associate Vice Provost for Curriculum, SUNY at Stony Brook.
10. F.D. Hutchinson, President, Dravo Energy Resources.
11. Clinton C. Kemp, Consultant, American Refuel, Canruf Company, Canada
12. Lee Koppelman, Executive Director, Long Island Regional Planning Board.
13. Evan Liblit, U.S. Environmental Protection Agency, Region 2.
14. Anthony Licata, Vice President, Dravo Energy Resources.
15. Parker D. Mathusa, Program Director, Energy Resources & Environmental Research, New York State Energy Research Development Authority.
16. Judith McEvoy, Assistant to the Director of Legislative and Economic Affairs, Long Island Association.

17. Homer A. Neal, Provost, SUNY at Stony Brook.

18. Linda O'Leary, Project Manager, Regional Waste Task Force, Port Authority of New York and New Jersey.

19. George Proios, Senate Executive Director, New York State Legislative Commission on Water Resource Needs of Long Island.

20. Frank Roethel, Associate Professor, Nassau Community College and Research Professor, MSRC, SUNY at Stony Brook.

21. Pat Roth, Ombudsman (Community Relations Specialist), New York State Department of Health.

22. T. Sanford, Regional Engineer, BFI of New York.

23. J.R. Schubel, Director, MSRC, SUNY at Stony Brook.

24. Ronald Scrudato, Research Associate, Rockefeller Institute of Government.

25. Frederick Seitz, President Emeritus, Rockefeller University.

26. Garrett Smith, Special Assistant for Air and Waste Management, U.S. Environmental Protection Agency, Region 2.

27. Marion R. Surgi, Signal Research Center

28. David Sussman, Vice President of Ogden Projects, Ogden Martin Systems, Inc.

29. A. Szurgot, Signal Environmental Systems.

30. Vincent Taldone, Office of Resource Recovery, New York City Department of Sanitation.

31. Peter M.J. Woodhead, Research Professor, MSRC, SUNY at Stony Brook.

32. Roberta Weisbrod, Special Assistant to Commissioner, New York State Department of Environmental Conservation.

# APPENDIX 4

## ENERGY AND MATERIALS RECOVERY FACILITIES[1]

| Location | Process | Products | Capac. (tpd)* | Capit. cost($M) | Status |
|---|---|---|---|---|---|
| **ALABAMA** | | | | | |
| Huntsville (Redstone Arsenal) | Mass burning in modular incinerator | Steam for heating & process | 50 | 3.2 | Shutdown 2/85 due to conveyor problems |
| Tuscaloosa | Mass burning in modular incinerator | Steam for process & heating by B.F. Goodrich Co. | D-300<br>T-300 | 8.5 | Operational since 2/84 |
| **ALASKA** | | | | | |
| Sitka | Mass burning of MSW and sewage sludge in modular combustion unit | Steam for heating use at Sheldon Jackson College | 25 | 4.2 | Operational since 5/85 |
| **ARKANSAS** | | | | | |
| Batesville | Mass burning in modular incinerator | Steam | D-50<br>T-55 | 1.2 | Operational since 5/81 |
| North Little Rock | Mass burning in modular incinerator | Steam for use by Koppers Co. (wood treating) | D-100<br>T-100 | 1.45 | Operational since 9/77 |
| Osceola | Mass burning in modular incinerator | Steam for heating and process at Crompton Osceola Co. | D-50<br>T-48 | 1.2 | Operational since 1/80 |
| **CALIFORNIA** | | | | | |
| City of Commerce | Mass burning in water-wall incinerator | Electricity for sale to Southern California Edison | 300 | 35.0 | Under construction since 3/85; startup expected in 3/87 |
| Fremont | Mass burning in modular incinerator | Electricity for sale to Pacific Gas & Electric | 480 | 35.0 | Air quality permit in public comment period; construction expected to begin in early 1986 with startup in late 1988 startup in late 1988 |
| Susanville | Mass burning of municipal waste & wood chips; electricity generation; hot water production | Electricity sold to PG&E; hot water for college's district heating system | 96 | 4.1 | Construction completed; commercial operation in 2/85 |

| Location | Process | Products | Capac.(tpd)* | Capit. cost($M) | Status |
|---|---|---|---|---|---|
| San Diego | Mass burn water-wall incinerator | Electricity | 2,250 | 227 | Applying for licensing with California Energy Commission. Site selected and approved. |
| Ukiah | Mass burning in modular incinerator | Electricity for sale to Pacific Gas & Electric Co. | 100 | 4.8 | Bonds issued; energy purchase agreement signed; constr. expected to begin in 10/85 with operation in 2/87. in 2/87 |
| CONNECTICUT | | | | | |
| Mid Conn. | RDF | Steam | 2,000 | 146 | Operational in 1988 |
| New Haven | | Electricity | 450 | 24 | Construction to begin 4/86; startup expected 8/88 |
| Wallingford | | Steam and Electricity | 420 | 25 | Startup scheduled for winter 1987 |
| Windham | Mass burning in modular incinerator | Steam; Electricity | D-108<br>T-125 | 7.0 | Operational since 1981; steam was used by Kendall Co., which closed in 1983; turbine generators now producing electricity. |
| DELAWARE | | | | | |
| Wilmington | Shredding, air classification, magnetic separation, froth flotation, other mechanical separation; aerobic digestion | RDF, ferrous & nonferrous metals; glass, humus | 1,000 tpd MSW coprocessed with 350 tpd of 20% solids digested sewage sludge | 72.3 | Solid waste processing in full operation since 3/84 |
| FLORIDA | | | | | |
| Broward Co. | Mass burning in water wall furnace for generation of electricity | Electricity for sale to Florida Power & Light | Southern Facility -2250; Northern Facility -2200 | 350.0 total for both | Construction expected to begin in early 1986 with operations in 1989 |
| Dade County | Hydrasposal (wet pulping), magnetic and other mechanical | Electricity for sale to utility, ferrous metals, aluminum & ferrous metals | D-3000<br>T-3000 | 165 | Operational since 1/82 |
| Hillsborough County | Mass burning | Electricity for sale to Tampa Electric Co. Co. | 1200 | 80.1 | Under construction since 1/85; startup expected in 7/87 |

| *Location* | *Process* | *Products* | *Capac.(tpd)** | *Capit. cost($M)* | *Status* |
|---|---|---|---|---|---|
| Lakeland | Shredding, magnetic separation, air classification, burning RDF with coal | Steam to produce electricity for use by City of Lakeland and Orlando Utility Commission, ferrous metals | D-300 T-200 | 5 for waste proc. plant | Operational, processing all of Lakeland's MSW ( ca 150 tpd) |
| Mayport Naval Station | Mass burning | Steam for use by base and ships | D-2 TPH T-120 tons per week (5 days) | 2.3 | Operational |
| Orange Co. (Walt Disney World) | Slagging pyrolysis incineration (Andco-Torrax) | High temp. hot water for heating and cooling at Walt Disney World | 100 | 15 | Plant is shut down owner is attempting to find interested party for further research & develop. of process |
| Panama City | Mass burning in rotary combustor | Steam and electricity | 510 | 38 | Under construction; operation expected 5/87 |
| Pinellas County | Mass burning, mechanical separation of metals after burning | Electricity for use by Fla. Power Corp., ferrous & nonferrous metals | D-2000 T-1842 | 80 | Fully operational since 5/83; expansion to 2764 tpd now underway, with compl. sched. for late 1986. |
| Pompano Beach | Shredding, magnetic and other mechanical separation, anerobic digestion of light fraction with sewage sludge | Methane gas, carbon dioxide | 50-100 | 3.65 | Operational (demonstration plant) |
| Tampa | Mass burning | Electricity to be sold to Tampa Electric Co. | 1000 | 59.9 (in 1981 $) | Operations began 9/85 |
| **GEORGIA** | | | | | |
| Savannah | Mass burning with modified water wall system | Steam and electricity for industrial use | 500 | 35 | Construction began 4/85 with operation expected in 4/87 |
| **HAWAII** | | | | | |
| Honolulu | Firing of RDF for generation of steam or electricity | Steam or electricity | 1800 | 145 | General obligaation bonds to be sold 8/85 |
| **IDAHO** | | | | | |
| Burley | Mass burning in modular incinerator | Steam for J.R. Simplot Co. (potato processing) | D-50 T-50 | 1.5 | Operational since 1/82 |

| *Location* | *Process* | *Products* | *Capac.(tpd)** | *Capit. cost($M)* | *Status* |
|---|---|---|---|---|---|
| **ILLINOIS** | | | | | |
| Chicago (Northwest Waste-to-Energy Facility) | Mass burning in water wall incinerators | Steam for process use on-site and by Brach Canndy Co. | D-1600<br>T-1250 | 23 | Operational |
| Chicago (Southwest Suppl. Fuel Process. Facility) | Shredding, air classification magnetic separation | RDF for use by utility; ferrous metals | 1000 | 19 | Published RFP in 2/84 for private sector to lease and operate |
| **IOWA** | | | | | |
| Ames | Baling waste paper, shredding, magnetic separation, air classification, screening, other mechanical separ-ation | RDF for use by utility, baled paper, ferrous metals | D-200<br>T-180 | 6.3 | Operational since 9/75 |
| **KENTUCKY** | | | | | |
| Campbells-ville | Mass burning in modular combustion units | Steam for process use by Union Underwear Co. | 100 | 4 | Project on hold for additional study |
| Ft. Knox | Mass burning in modular incinerator | Steam for heating & air conditioning at hospital | 40 | 1.9 | Construction complet-ed, but modifications needed before full-scale operations can begin. |
| **LOUISIANA** | | | | | |
| New Orleans | Shredding, air classification, magnetic and other mechanical separ-ation | Ferrous metals | D-770<br>T-650 | 9.1 | Shredding/land-filling and ferrous recovery operational |
| **MAINE** | | | | | |
| Auburn | Mass burning in modular incinerator | Steam for heat and process at Pioneer Plastics | D-200<br>T-170 | 3.98 | Operational since 4/81 |

| *Location* | *Process* | *Products* | *Capac.(tpd)** | *Capit. cost($M)* | *Status* |
|---|---|---|---|---|---|
| **MARYLAND** | | | | | |
| Baltimore (Southwest Resource Recovery Facility) | Mass burning in water-wall furnace; electricity generation, ferrous recovery from ash | Electricity for sale to Baltimore Gas & Electric Co; ferrous metals; plannned sale of steam to district heating system. | 2250 | 170 | Operational since 5/85 |
| Baltimore County | Shredding, magnetic and other mechanical separation | RDF, ferrous metals, glass nonferrous metals | D-1200 T-850 | 11.0 | Operational; recovering ferrous metals and glass, producing utility grade ash; shredded RDF used in BG&E Co. cyclone |
| Harford Co. | Mass burning in modular combustion units | Steam for space heating & process use by U.S. Army at Aberdeen Proving Ground | 300 | 14 | Construction to begin in late 1985. Operation expected by 6/87 |
| **MASSACHUSETTS** | | | | | |
| Haverhill & Lawrence | Shredding, magnetic separation, trommel screening at Haverhill; burning RDF for cogeneration of steam and elect. in Lawrence | Steam and electricity for industrial use; surplus electricity sold to utility | 1300 | 99.5 | Started commercial operation in 9/84 |
| Millbury (Central MA Resource Recovery | Mass burning in water wall boilers, electricity generation | Electricity for sale to local utility | 1500 | 150 | Waste disposal contracts signed; permitting in progress; construction began in 7/85, with startup planned for 12/87 |
| North Andover | Mass burning in water wall furnace, electricity generation | Electricity for sale to utility | D-1500 T-1500 | 196 | Commercial operation expected by late 1985 |
| Pittsfield | Mass burning in modular incinerator | Steam for process & heating by Crane & Co. | D-240 T-240 | 10.8 | Operational since 3/81 |
| Rochester | Shredding, magnetic separation, burning PRF in stoker-grate boiler, non-ferrous recovery from ash, generation of electricity | Electricity for sale to Commonwealth Electric; ferrous and nonferrous metals | 1500 | 160 | Bonds sold 12/84; construction expected to begin in late 1985, contingent upon obtaining air quality permit |
| Saugus | Mass burning in water wall furnaces, magnetic separation | Electricity for sale to utility; ferrous metals | D-1500 T-1200 | 50 | Operational; conversion to electric power generation completed 9/85 |

| Location | Process | Products | Capac.(tpd)* | Capit. cost($M) | Status |
|---|---|---|---|---|---|
| **MICHIGAN** | | | | | |
| Detroit | Flail milling, trommel screening, secondary shredding ,burning RDF in on-site boilers burning RDF, elec. generation in 65MW turbo-generator | Steam for Detroit Edison's central heating system; electricity for sale to Detroit Edison; ferrous metals. | 3300 | 200 | Negotiating with Combustion Eng- prior to contract signing; all permits in hand; remarketing of bonds in process |
| **MINNESOTA** | | | | | |
| Collegeville | Mass burning in modular incinerator | Steam for heating, electricity generation & other uses by university. | D-58<br>T-43 | 2.4 | Operational since 11/81 |
| Duluth | RDF process rebuilt includes primary disk screen, shredding, air knife, sizing disk screen, fluidized bed incineration of RDF and sludge | RDF, ferrous metals, steam for heating and cooling of plant and to run process equipment | 400 tons of MSW per shift; 340 of 20% solids sewage sludge | 19 | Operational |
| Newport | Production of refuse-derived fuel; burning to produce elec; separation of ferrous metals and aluminum | Electricity, ferrous metals, aluminum | 1000 | 20.75 | Construction began in 7/85, operation expected by 7/87 |
| Red Wing | Mass burning in modular incinerator | Steam for S.B. Foot Tanning Co. | 72 | 2.5 | Operational since 9/82 |
| **MISSISSIPPI** | | | | | |
| Pascagoula | Mass burning in modular combustion unit | Steam for process use by Morton Thiokol | 150 | 6.0 | Operational as of 1/85 |
| **MISSOURI** | | | | | |
| Ft. Leonard Wood | Mass burning in modular incinerator | Steam for cooking and heating in barracks complex | D-75<br>T-30 | 3.3 (approx.) | Operational |
| St. Louis | Mass burning, electricity generation | Steam for downtown district heating & cooling electricity for sale to Union Electric Co. | 900 | 30-40 | District heating and cooling purchased by Thermal Resources and Bi-State; final stage of contract negotiations for waste plant; start up expected in Fall 1987 |

| *Location* | *Process* | *Products* | *Capac.(tpd)** | *Capit. cost($M)* | *Status* |
|---|---|---|---|---|---|
| **MONTANA** | | | | | |
| Livingston | Mass burning in modular incinerator | Steam for heating at Burlington Northern Railroad repair shops | D-75<br>T-70 | 2.6 | Operational since 5/82 |
| **NEVADA** | | | | | |
| Reno | Processing to remove glass & metals to produce RDF; carbon char (K- Fuel) by pyrolysis; burning in fluidized bed combustor; electr. generation | Electricity for sale to utility, glass, ferrous metals, and aluminum recovered | 250<br>Phase I<br><br>1000<br>Phase II | 50 | Phase I operating; Phase II construction expected to begin in early 1986. |
| **NEW HAMPSHIRE** | | | | | |
| Claremont (NH/VT Solid Waste Project) | Mass burning in water wall boiler | Electricity for sale to Central Vt. Public Service Co. | 200 | 17.9 | Construction underway; operation expected in 5/87 |
| Durham (Lamprey Regional Solid Waste Cooperative) | Mass burning in modular incinerator | Steam for heating & hot water at Univ. of N.H. | D-108<br>T-100 | 3.3 | Operational since 9/80 |
| Groveton | Mass burning in modular incinerator | Steam for industrial use | 24 | n/a | Operational |
| Manchester | Mass burning in modular incinerator | Electricity for sale to Public Service of New Hampshire | 450 | 20.0 | Approximately twelve adjacent communities have agreed to participate in project; construction to start in early 1986, with startup in Spring 1988 |
| Portsmouth | Mass burning in modular incinerator | Steam for heating at Pease AFB | D-200<br>T-200 | 6.25 | Operational since 7/82 |
| **NEW JERSEY** | | | | | |
| Bergen County | Mass burning in water wall furnace for generation of electricity | Electricity for sale to Public Service Electric & Gas | 3000 | 253 | Financing complete; construction to begin mid-to-late 1985, with operations in 1988 |

| Location | Process | Products | Capac.(tpd)* | Capit. cost($M) | Status |
|---|---|---|---|---|---|
| Essex County | Mass burning for electricity generation | Electricity for sale to utility | 2250 | 200 | Negotiating contracts with BFI; construction scheduled for Fall 1985 with start-up in summer 1988; will include intermediate processing facility on-site for source separated materials |
| Ft. Dix | Mass burning of presorted solid waste in modular incinerator | Steam for heating on base | D-80<br>T-60 | 6 | Contract awarded to joint venture firm. Under construction since 7/85; operation expected in 7/86. |
| Warren Co. | Mass burn water wall incinerator | Electricity | 400 | 40.3 | Construction to begin 5/86 startup expected 5/88 |
| NEW YORK | | | | | |
| Albany | Processing plant; shredding, magnetic separation. Burning PRF in stoker-grate boiler; ash processing center; ferrous, & aggregate recovery from boiler ash | Processed refuse fuel (PRF), steam for heating and cooling state offices, ferrous & nonferrous metals, boiler aggregate | D-750 tons per shift<br>T-750 per shift | 28.2 | Operational |
| Brooklyn Navy Yard) NYC | Mass burn | Steam and electricity; ferrous metal | 3000 | 290 | Operation expected in 1989 |
| Cuba (Cattaraugus Co. Refuse-to-Energy Facility) | Mass burning in modular incinerator | Steam for process at Cuba Cheese Co. | D-112<br>T-120 | 5.5 | Operational since 2/83 |
| Dutchess Co. | Mass burning in O'Connor rotary combustor for generation of steam and electrcity; ferrous metals recovery | Steam for sale to IBM Corp., electricity to utility, ferrous metals | 400 | 30 | Project financed in 12/84; under construction since 12/84; operation expected in early 1987 |
| Glen Cove | Mass burning in stoker-fired furnace with centrifuges sewage sludge | Electricity for sewage treatment plant and incinerator; excess to Long Island Lighting Company. | 250 | 34 | Operational since 8/83 |
| Hempstead | Mass burning; electricity generation | Electricity for sale to utility | 2250 | 250 | Negotiations underway between Town and American REF-FUEL; financing expected by 1986. Construction expected in early 1986, with operation in 33 months. |

| *Location* | *Process* | *Products* | *Capac.(tpd)** | *Capit. cost($M)* | *Status* |
|---|---|---|---|---|---|
| Islip (MacArthur Energy Recovery Facility) | Mass burning | Electricity for sale to Long Island Lighting Company; ferrous recovery from ash. | D-518 | 39.5 | Bond sale in 9/85; expected startup in late 1987. |
| Monroe Co. | Shredding, air classification froth flotation, magnetic and other separation | RDF for use by utility as supplemental boiler fuel, ferrous metals, glass | D-2000<br>T-400 | 62.2 | Facility closed 7/27/84; currently preparing RFP for alternative use |
| New York (Betts Ave. Incinerator) | Mass burning in refractory furnace | Steam for heating and processes in-plant and adjacent City garages | 1000 | 5 | Closed for design review and possible renovation |
| Niagara Falls | Shredding, magnetic separation, burning shredded refuse | Steam for use by chemical plant; electricity sold to power company grid; ferrous metals | D-2000<br>T-1700 | 100 + | Operational |
| Oneida County | Mass burning in modular combustion units | Steam for heating, hot water & other use by Griffis Air Force Base; electricity from excess steam | 200 | 13.5 | Operational since 1/85; testing conducted in 3/85; turbine being added to generate electricity from summer excess steam |
| Oswego County | Mass burning in modular combustion units, electricity generation | Steam for use by Armstrong World Enterprises; electricity for sale to Niagara Mohawk | 200 | 14.5 | Under construction; startup expected in 10/85 |
| Oyster Bay | Mass burning, electricity generation | Electricity for Long Island Lighting Co. | 1650 | 113 | Contract negotiations under way; construction expected to begin in 1986 with operation in 1989 |
| Washington County | Mass burning, production of electricity | Electricity for sale to utility | 240 | 48 | Energy purchase agreement signed; bonds sold; construction pending environmental litigation and N.Y. State construction permits |

| *Location* | *Process* | *Products* | *Capac.(tpd)** | *Capit. cost($M)* | *Status* |
|---|---|---|---|---|---|
| Westchester County (Peekskill) | Mass burning in water wall furnace, electrical generation, ferrous metal recovery from ash | Electricity for Consolidated Edison Co., ferrous metals | 2250 (Permitted capacity 1980) | 179 | Began startup 2/84; commercial operation 10/84 |
| NORTH CAROLINA | | | | | |
| New Hanover County | Mass burning in water wall boilers, cogeneration of steam and electricity | Steam for use by W.R. Grace Co. (agrochemical mfr.); electricity for sale to Carolina Power & Light | 200 | 13 (approx.) | Operational |
| NORTH DAKOTA | | | | | |
| Williston | Mass burning; cogeneration of steam & electricity | Steam for process use by Hardy Salt; electricity for sale to utility | 100 | 5 | Awaiting final energy contracts; groundbreaking expected in 1986 |
| OHIO | | | | | |
| Akron | Shredding, magnetic separation, burning RDF in semi-suspension stoker-grate boiler | Steam for urban and industrial heating and cooling, ferrous metals, hot water for residential and commercial heating | D-1000 T-900 | 80 | Temporarily closed while repairing damage from explosion in 12/84; natural gas being used to provide steam to customers |
| Columbus | Shredding, magnetic separation, burning of shredded refuse with supplemental coal in semi-suspension stoker-grate boiler to produce steam and generate electricity | Electricity for city customers | D-2000 T-1500 | 175 | All units operational; making modifications to boilers & support systems |
| Gahanna | Magnetic separation, disk & trommel screening, shredding, densification of RDF, composting | Organi-FUEL 100[tm] (d-RDF) for gasification in oil or gas boilers and use as supplemental fuel; compost, aluminum, glass, paper, ferrous metals, other recyclable materials | D-1000 T-400 | 10 | Operational since 11/81; plant also has buyback center for recyclable materials |

| *Location* | *Process* | *Products* | *Capac.(tpd)** | *Capit. cost($M)* | *Status* |
|---|---|---|---|---|---|
| Montgomery County | Renovation of an existing incinerator | Steam | 300 | 8.6 (for rehab) | Renovation to be completed 5/87 |
| **OKLAHOMA** | | | | | |
| Miami | Mass burning in modular incinerator | Steam for industrial use by B.F. Goodrich Co. | D-108 T-72 | 3.14 | Operational since 11/82 |
| Oklahoma City | Phase I shredding, ferrous & nonferrous metals separation; burning in rotary drum furnace and electricity generation. Phase II has anerobic digestion of organic MSW and sewage | Electricity & methane gas for sale to Okla. Gas & Electric Co.; ferrous & non-ferrous metals | 5600 tons per week (Phases I & II) | 29 | Phase I startup testing com-pleted; con-tinuous opera-tion expected to begin in mid-1985; awaiting decision on anerobic diges-tion vs. thermal reduction for Phase II |
| Tulsa | Mass burning, generation of steam and electricity | Steam for sale to Sun Refin-ing; electri-city for sale to Public Service Co. of Okla. | 750 | 51.5 | Groundbreaking in 5/84; startup expected in 1/86 |
| **OREGON** | | | | | |
| Marion County | Mass burning in water wall furnaces | Electricity for local utility | 550 | 47.5 | Under construc-tion; startup expected in Spring 1986 |
| **PENNSYLVANIA** | | | | | |
| Delaware Co. | Mass burning in modular incinerator | Steam for use by Fair Acres Geriatrics Inst | 50 | 2.9 | Groundbreaking in 12/84; con-struction on hold pending zoning decision |
| Erie | Shredding, mechanical separation, air classifi-cation, burn-ing RDF, elec-tricity gene-ration | RDF for use as fuel to produce elec-tricity for local utility steam, ferrous metals, glass | 600 | 30 | In design and permitting stage; startup expected in 3/87 |

| Location | Process | Products | Capac. (tpd)* | Capit. cost($M) | Status |
|---|---|---|---|---|---|
| Harrisburg | Mass burning of MSW and sewage sludge in water wall furnace, bulky waste shredding | Steam for utility-owned district heating system and for city-owned sludge drying system, excess steam to Bethelem Steel, ferrous metals | D-720 T-700 | 8.3 | Operational since 1973; sludge drying facility in test; overhauling plant to reestablish process reliability. to be on-line in 5/86. |
| **RHODE ISLAND** | | | | | |
| Quonset Industrial Park | Mass burning in water wall furnace | Steam and electricity | 1500 | 100 | Project financed; construction to begin in Spring 1986. |
| **SOUTH CAROLINA** | | | | | |
| Johnsonville (Wellman Energy Plant) | Mass burning of industrial waste in modular incinerator | Steam for process use by Wellman Industries | D-50 T-50 | 2.5 | Operational since 1981 |
| **TENNESSEE** | | | | | |
| Dyersburg | Mass burning in modular incinerator | Steam for process & heat at Colonial Rubber Works | D-100 T-82 | 2 | Operational since 9/80 |
| Gallatin | Mass burning in water wall rotary combustor for cogeneration of steam & electricity; PREBURN™ materials recovery system | Steam for industrial processing and electricity for sale to TVA, ferrous metals and aluminum | 200 | 10 | Operational since 12/81 |
| Lewisburg | Mass burning in modular incinerator | Steam for industrial use by Heil-Quaker Corp. | D-60 T-35-40 | 1.75 | Operational since 1980 |
| Nashville | Mass burning in water wall incinerator | Steam and chilled water for urban heating and cooling; expansion adds electricity for sale to TVA | D-720 T-612 | 24.5 | Operational since 1974; expansion to be completed in early 1986, increasing design capacity to 1120 |

| *Location* | *Process* | *Products* | *Capac.(tpd)** | *Capit. cost($M)* | *Status* |
|---|---|---|---|---|---|
| **TEXAS** | | | | | |
| Cleburne | Mass burning in modular combustion unit | Steam or electricity | 115 | 5.5 | Under construction, with completion expected in 10/85 |
| Galveston | PREBURN™ materials recovery system | Ferrous & metals aluminum | 200 | 1.1 | Under contract; operations expected 12/85 |
| Lubbock | Mass burning in water wall rotary combustor for generation of electricity; PREBURN™ materials recovery system | Electricity for sale to Lubbock Power & Light ferrous metals & aluminum | 500 | 42 | Contract awarded operations expected 12/87 |
| Waxahachie | Mass burning in modular incinerator | Steam for industrial use by International Aluminum Extruders | D-50 | 2.2 | Operational since 7/82; selling 60% of steam produced |
| **VERMONT** | | | | | |
| Rutland | Mass burning in modular incinerator, electricity generation | Electricity for sale to Central Vermont Public Services Corp. | 240 | 17 | Construction began 10/85; completion expected in Spring 1987 |
| **VIRGINIA** | | | | | |
| Alexandria/ Arlington | Mass burning in water wall incinerator | Electricity for sale to Virginia Power Co. | 975 | 54.1 | Construction began 5/85; startup expected 8/87 |
| Galax | Mass burning in rotary combuster for generation of steam | Steam for sale to Hanes Knitwear, Inc. | 55 | 2.1 | Under construction since 5/84; startup expected in 9/85 |
| Hampton | Mass Burning in water wall furnace | Steam for use by NASA Langley Research Center | D-200<br>T-200 | 10.4 | Operational since 9/80 |
| Harrisonburg | Mass burning | Steam for heating & cooling at James Madison Univ. | D-100<br>T-75 | 8 | Operational since 12/82 |
| Newport News (Ft. Eustis) | Mass burning in modular incinerator | Steam for heating, hot water & cooking | D-40<br>T-30 + | 1.7 | Operational since 12/80 |

| *Location* | *Process* | *Products* | *Capac.(tpd)** | *Capit. cost($M)* | *Status* |
|---|---|---|---|---|---|
| Norfolk (Norfolk Naval Station) | Mass burning in water wall furnace | Steam for use by facilities at Norfolk Naval Station | 360 (two 180 tpd boilers operated alernately) | 2.2 (1967) | Operational; temporarily shut down to retube boilers |
| Petersburg | Phase I- shredding, magnetic and other separation, sale of RDF cubes; Phase II- addition of boiler/turbine to generate electricity; Phase III- ethanol production, 10 million gal/yr; Phase IV- possible location for 50 tpd cellulose/alcohol R&D facility; Phase V- 37.5 million gal/yr cellulose/alcohol production based on enzymatic hydrolysis process | Phase I & II- ferrous and nonferrous metals, glass electricity for sale to utility, steam for in-plant use; Phases III, IV & V- ethanol, $CO_2$, dried grain supplement (DGS), distiller's dried grain supplement (DDGS) | 2000 (peak) 650 initial to 2400 tpd with backup fuel of wood chips and agricultural waste | 12 (Phase I) 11 (Phase II) 7 (Phase III) 33 (Phase IV) 136 (Phase V) 200 Total | Preliminary design completed groundbreaking expected in Summer 1985 with startup 9 months later for Phase I |
| Portsmouth (Norfolk Navy Shipyard) | Mass burning in water wall furnace | Steam for use by facilities at Naval Shipyard | 160 (two 80-tpd boilers, operated alternately) | 4.5 | Operational |
| Portsmouth (Southeastern Tidewater Energy Project) | Shredding, magnetic and other separation | RDF for burning in new RDF/coal augmented power plant under construction at Naval Shipyard, proving steam and electricity; ferrous and nonferrous metals. | 2000 | 50 | Under construction; operations expected in Spring 1987 |
| Richmond | Flail milling, magnetic separation, disc screening, densification of RDF, hand-sorting aluminum | Densified RDF; ferrous metals; aluminum; compost material | 250 | 3.2 | Operational |
| Salem | Mass burning in modular incinerator | Steam | 100 | 1.9 | Operational since late 1970s |
| **WASHINGTON** | | | | | |
| Tacoma | Shredding, air classification magnetic separation | RDF, ferrous metals | 500 | 2.5 | RDF plant closed; city is evaluating bids received 3/84 for a new waste-to-energy facility |

| Location | Process | Products | Capac.(tpd)* | Capit. cost($M) | Status |
|---|---|---|---|---|---|
| **WISCONSIN** | | | | | |
| Madison | Shredding, magnetic separation, trommel screening, secondary shredding | RDF burned with coal at Madison Gas & Elec. Co. for electrcity generation; RDF burned with coal at Oscar Mayer Foods Corp. for steam production; ferrous metals. | D-400 T-250 | 2.5 | Refuse processing & burning at Madison Gas & Electric operational since 1/79; Oscar Mayer installation operational since 6/83 |
| Waukesha | Mass burning in refractory furnace | Steam for local industry and sewage treatment plant | D-175 T-140 | Incinerator 1.7 (1971) Heat recovery system 3.9 (1979) | Incinerator operating since 1971; waste heat recovery boiler added in 1979; operating and sending steam to local industry and sewage plant |
| **CANADA** | | | | | |
| **ONTARIO** | | | | | |
| Hamilton | Shredding, magnetic separation, semi-suspension burning in dedicated speader stoker boilers | Electricity for Ontario Hydro, steam for in-plant use, ferrous metal | D-500 T-450 | 9 + (1972) | Operational since 1972; 4.0 MW turbine generator added and operating since 11/82; $12 million modernization program to be completed by Summer 1986 |
| Toronto | Shredding, air classification, screening, mass burning in modular incinerator with heat recovery. | Ferrous metal, RDF compost; hot water for plant heating | Resource recovery-220; transfer facility-600 | 15 + | Operational since 3/77 |
| **PRINCE EDWARD ISLAND** | | | | | |
| Parkdale | Mass burning in modular incinerator | Steam for heating/cooling at hospital complex | 108 | 8.2 + since 2/83 | Operational |

| *Location* | *Process* | *Products* | *Capac.(tpd)** | *Capit. cost($M)* | *Status* |
|---|---|---|---|---|---|
| **QUEBEC** | | | | | |
| Montreal | Mass burning in water wall furnaces<br><br>customers | Steam used by City offices & facilities and private | D-1200<br>T-1200 | 14.7 + (1967) | Operational since 1970; gas cleaning systems redesigned and and rebuilt in 1983; steam sold to 20 customers through 10 miles of underground pipeline |
| Quebec | Mass burning in water wall furnace | Steam, used for industrial process by paper mill | 1000 | 25 + (1974) | Operational since 1974 |

*D = Design Capacity; T = Actual throughput (recent average)

[1]After Waste Age November 1985 and the U.S. Conference of Mayors

# GLOSSARY

**Acid Rain.** Rainfall of lowered pH (acidified) due to absorption of sulfur dioxide, nitrogen oxides, and certain other pollutants from the air. Sulfur dioxide and nitrogen oxides form sulfuric and nitric acid respectively.

**Activated Charcoal.** A form of carbon frequently used in air and/or water filters to remove organic contaminants.

**Air Classification.** A process of separating light from heavy shredded wastes in an injected stream of air within a controlled chamber.

**Aquifer.** An underground layer of porous rock, sand, or other material in which infiltrated water collects and flows between layers of non-porous rock or clay. Aquifers are frequently tapped for wells.

**Appropriate Technology (intermediate or low technology.)** Technology of low capitalization and overhead, that is matched to local material and human resources, designed to produce goods and services for local consumption.

**Back End System.** That portion of a High Technology Resource Recovery Facility where reclaimable materials are extracted from the residues of incinerated refuse.

**Back Haul.** Low cost transport of secondary material (refuse) resulting from prior underutilization of transfer vehicle volume capacity.

**Bacteria.** Any of numerous species of simple, single-celled, microscopic organisms that reproduce by simple cell division. Along with fungi, bacteria comprise the decomposer component of ecosystems. Some species are agents of disease.

**Baghouse.** A pollution control device which uses a filter bag to dissociate and collect particulates from a gas stream -- similar in operation to a vacuum cleaner bag.

**Benthic Fauna.** The group of organisms which colonize sediments beneath a body of water. These can be subdivided into Epifauna - those living on the surface of the sediments and Infauna - those living within the sediments.

**Bioaccumulation.** The accentuated accumulation of potentially toxic substances in organisms at successive trophic levels. Due to excretory or biodegradatory incapacity, absorbed or ingested heavy metals and chlorinated hydrocarbons may concentrate a million-fold in organisms high on the food chain. Also called biomagnification.

**Bioconversion.** The conversion of organic waste via biological decomposition (mediated by bacteria and fungi) to produce usable natural gas, liquid fuels or compost products.

**Biodegradable.** Able to be consumed and broken down to natural substances such as carbon dioxide and water, particularly by decomposer organisms.

**Biological Oxygen Demand (BOD).** A measure of water quality; the amount of dissolved oxygen required by biodegrader organisms to decompose organic matter in a given volume of water. Greater BOD indicates a poorer water quality.

**Boiler Blowdown.** Wastewater, discharged from lower boiler drums, in which the precipitates of chemicals introduced during the treatment of boiler feed water are collected.

**Bottom Ash.** The residual solids on a hearth or grate following incineration of wastes.

**Buy-Back Center.** A facility at which reclaimed waste material is purchased from resource suppliers to be reintroduced to the commodity market in its raw or processed form.

**Cap/Cover.** Cover is soil applied over solid wastes at the end of each workday in an landfill. A Cap is a permanent layer of impervious material (clay, polyethylene or PVC) added to the Cover at the completion of the landfill.

**Carbon Monoxide (CO).** An odorless, colorless, toxic gas produced in the incomplete combustion of carbonaceous fuels.

**Carrying Capacity.** The maximum population of an organism that a given habitat can support.

**Cellulose.** A carbohydrate (a substance containing carbon, hydrogen and oxygen) that is a major component of the cell walls of many plants, including trees and vegetables. Paper is almost totally cellulose. Wood is approximately 40 percent cellulose.

**Chemical Oxygen Demand (COD).** A measure of the chemical content of water; the amount of oxygen required, per unit time, for oxidation reactions involving chemicals in solution.

**Chlorinated Hydrocarbons.** Synthetic organic molecules in which one or more hydrogen atoms are replaced by chlorine atoms. The compounds are nonbiodegradable and, therefore, a potential health hazard. Many have been shown to produce cancer in laboratory animals. Also called organochlorides.

**Classical Recycling.** The use of secondary materials, recovered waste resources, to produce new products.

**Clay Capping/Clay Lining.** The application of a clay layer cap over a completed landfill site to minimize leachate production by restricting groundwater infiltration. Some landfills are clay lined prior to the introduction of refuse.

**Cogeneration.** The production of both steam and electricity by a resource recovery facility.

**Collection Center.** A facility designed to accept and transfer trash commodities such as glass bottles, paper products and/or aluminum cans.

**Combustible Fraction.** Solid waste which can be incinerated with or without supplementary fuel.

**Combustion.** Burning of organic material in the pressence of oxygen.

**Compaction.** Packing down. Aeration and water infiltration are reduced in compacted soil, thereby limiting its capacity to support life. Compacted trash has reduced landfill space requirement.

**Composting.** Controlled disposal of solid organic wastes by biological decomposition to a state where the product, compost, is environmentally inert or beneficial. Compost may be used as a soil conditioner. Commercial composting is conducted out-of-doors in windrows, or in mechanical aeration tanks within resource recovery plants.

**Criteria Pollutants.** Those airborne chemical pollutants used to define the National Ambient Air Quality standard. In accordance with the Clean Air Act, these include carbon monoxide, ozone, lead, sulfur dioxide, nitrogen oxides and hydrocarbons.

**Cullet.** Mixed glass fragments.

**Decomposition (anaerobic).** The biological breakdown of refuse to simple inorganic constituents (e.g. hydrogen sulfide and methane) by bacteria and fungi in the absence of free oxygen.

**Dioxin.** A synthetic organic compound, of the chlorinated hydrocarbon class (polychlorinated dibenzo dioxins, PCDDs) known to variously cause birth defects, skin disorders, liver damage, immune system suppression and cancer in laboratory animals at very low doses. The exact sensitivity of humans is not known. Dioxin is produced in the combustion of solid wastes and in the manufacturing of some herbicides and wood preservatives. It has become a widespread environmental pollutant of which the most toxic form is 2,3,7,8 tetrachlorodibenzodioxin.

**Dry Scrubber.** A pollution control device which sequesters particulate and gaseous impurities in exhaust emissions to an injected scurried reagent within the exhaust flue of combustion chambers.

**Dump.** An area in which solid waste is disposed but not daily covered with soil, thereby not meeting the criteria of a sanitary landfill.

**Eh (Oxidation-reduction potential.)** A measure of a particular environment's ability to oxidize (burn/decompose) organic material. Lower Eh indicates a poorer oxidizing ability.

**Effluents.** The gaseous or wastewater discharge to the environment from industry or sewage treatment plants.

**Electrostatic Precipitator (ESP).** A pollution control device which traps electrically charged particles out of an air emissions exhaust stream.

**Emission.** The release of gaseous combustion products into the atmosphere.

**Energy Recovery.** A form of resource recovery in which the organic fraction of waste is converted to usable energy. Energy recovery from processed or raw refuse may be achieved through combustion to produce steam (e.g. as supplemental fuel in electric utility power plant boilers); through pyrolysis to produce oil or gas; and through anaerobic digestion to produce methane gas.

**EPA (United States Environmental Protection Agency.)** The federal agency responsible for protecting the environment from all forms of degradation.

**Ferrous Metals.** Metals containing a high percentage of iron.

**Fly Ash.** Non-combustible particles carried by flue gas.

**Front End Recovery.** A waste treatment system which combines mechanical separation and resource recovery of paper fibers, glass, metals and compostable organics from raw, untreated, solid wastes prior to detailed processing. Solid waste volume may be reduced by 20 percent in the process.

**Furans (Polychlorinated dibenzofurans, PCDFs).** Molecules consisting of two benzene rings forming a five sided core in which one oxygen atom is contained. These compounds are very similar to dioxins, both in molecular structure and physiological effect.

**Garbaeology.** The scientific study of garbage for the purpose of adducing the relationships between material culture and human behavior. William Rathje, Univeristy of Arizona archeologist, developed the methodology as an inexpensive and inconspicuous method of sampling of populations which are difficult or impossible to interview.

**Garbage.** Biodegradable waste material.

**Grading.** Hand sorting of mixed heterogenous materials into homogenous categories.

**Groundwater.** Fresh water deposits, accumulated in subterranean aquifers and replenished by infiltrating rain, which serve as reservoirs for springs and wells.

**$H_2S$ (Hydrogen Sulfide).** A reduced inorganic compound produced in anaerobic digestion of organic matter by sulfate reducing bacteria.

**Halogenated Hydrocarbons.** Synthetic organic compounds containing one or more atoms of chlorine, fluorine and/or bromine, of the halogen group.

**Hazardous Waste.** Chemical or biological refuse, of industrial or consumer origin, considered potentially dangerous to humans and/or the environment as established by EPA standards.

**Heavy Metals.** Any of the high atomic weight metals such as lead, mercury, cadmium and zinc. All constitute a serious pollution threat because of their toxicity in relatively low concentrations and their tendency to bioaccumulate.

**Integrated Resource Recovery.** The reclamation of materials and energy from waste by different, yet compatible, sectors of society.

**$LC_{50}$** A measure of the toxicity of a chemical, in which a single dose is lethal to 50% of the animal population tested.

**Leachate.** The contaminated fluid resulting from water percolating through a waste landfill site.

**Liner (Impermeable).** A natural or man-made material (usually clay or PVC) used to seal the bottom and sides of sanitary landfills in order to contain leachate.

**Litter.** Haphazardly disposed consumer waste.

**Mass Burning.** Direct combustion of unprocessed refuse.

**Mass-Burning Waterwall Incineration.** Direct combustion of unprocessed refuse within a waterpipe jacketed furnace, resulting in energy recovery in the form of steam.

**Materials Recovery.** Extraction of waste commodities from solid waste by manual and/or mechanical means.

**Mechanical Separation/Recovery.** Mechanical (electromagnetic, air classification, etc.) segregation of mixed waste into homogeneous material categories for recovery.

**Metal (Ferrous).** Iron containing metals which commonly exhibit magnetic properties. Ferrous metal waste ("tin" cans, automobiles, etc.) can be electromagnetically separated from mixed refuse.

**Methane ($CH_4$).** A simple organic compound, existing as a gas at room temperature, which is produced in anaerobic digestion of refractory organic matter by methanogenic bacteria.

**Midden.** Arefuse mound, consisting of the discards of an earlier civilization (usually shells, pottery, bones, etc.), giving archaeologists and paleohistorians insights into the culture of that people.

**Mixed Waste.** Randomly associated waste products, by virtue of which the value of an individual product is reduced to a lowest common denominator.

**Municipal Solid Wastes.** Residential and nonindustrial garbage, trash, sludges and material discards.

**NEPA.** National Environmental Policy Act.

**NIMBY Syndrome (Not-In-My-Back-Yard).** Reactionary affirmation of communities (or individuals) to the potential introduction of solid waste disposal facilities (or other such operations) in their vicinity, resulting from a fear of a concommitant diminished quality of life.

**$NO_3$ (Nitrate).** A chemical compound containing nitrogen (an element essential to plant growth) in an oxidized form. Under anoxic conditions, $NO_3$ is reduced to $N_2$ by denitrifying bacteria; thus its presence indicates toxic conditions which may be correlated with inhibited plant growth.

**Nonbiodegradable.** Refers to substances (including plastics, metals and many synthetic chemicals) which cannot be chemically broken down to simple molecules (usually carbon dioxide and water) in biological decomposer food chains. Those nonbiodegradable toxic chemicals which tend to bioaccumulate are a particular environmental and health hazard.

**Nonburnable.** Materials that cannot be incinerated in a combustion chamber and must be disposed of with residual ash.

**Non-Ferrous Metals.** Metals and alloys containing either no iron or very small amounts as impurities or alloying additions. Nonferrous metal commodities may be composed of brass or elemental metals such as copper or aluminum.

**Nonrenewable Resources.** Various raw materials (including metal ores, oil and coal) of economic value existing as finite, nonreplenishable deposits in the earth's crust.

**Organic Waste (putrescibles).** Biodegradable solid refuse, characterized by the prevalence of carbon atoms in the material's molecular structural matrix (e.g. food scraps.) (Synonym: garbage).

**Organometallic.** A compound consisting of an organic molecule associated with a metal ion such as iron, copper, nickel or aluminum.

**Partial Feedstock.** Virgin or secondary material used in addition to primary feedstock in industrial processes.

**Particulates.** Partially incinerated microscopic matter (e.g. ash, soot, dust and charred paper) suspended and discharged in exhaust gases.

**Parts Per Million (ppm).** An expression of concentration relating the number of units of one substance to a million units of another. (e.g., one gram of $NO_3$ dissolved in a million grams of water has a concentration of 1 ppm).

**Pathogen.** A disease-causing organism.

**pH.** A measure of the acidity of a solution (water + chemical solute.) Strictly, pH is a measure of the hydrogen ion concentration and follows the formula, $pH = -\log_{10}[H^+]$.

**PCBs (Poly-chlorinated Biphenyls).** A specific group of synthetic compounds, in which chlorine is attached to two benzene rings.

**Plastics.** Polymers derived from petroleum which are increasingly used as substitutes for glass, metal, wood, leather, etc. Thermoset plastics are nonremeltable and irrecoverable; thermoplastics can be melted and reused.

**Primary Feedstock.** Virgin or secondary material used in the greatest amount in industrial processes.

**Pyrolysis.** The chemical decomposition of a material by heat, in the absence of oxygen, yielding a gaseous or liquid product which can be used as a fuel.

**RDF Dedicated Boiler.** A steam producing, waterwall combustion chamber in which shredded refuse derived fuel (RDF) is held in suspension (or semi-suspension) during burning. This technology uses front-end separation for recovery of noncombustible materials. (See Mass-Burning Waterwall Incineration).

**Reactive $PO_4$ (Phosphate).** A chemical compound containing phosphorous, an element essential for plant growth, in its most prevalent oxidized form.

**Recyclable.** Refers to a waste commodity which can be readily reutilized in a similar or altered form.

**Recycling.** The separation of reusable materials from waste followed by processing for repeated consumption.

**Refuse.** All organic putrescible and nonputrescible solid wastes (excepting body wastes,) including garbage, rubbish, dead animals and solid industrial wastes.

**Refuse Compaction Vehicles.** Collection vehicles equipped with hydraulic systems to compress voluminous waste materials en route to disposal sites.

**Refuse Derived Fuel (RDF).** An easily handled fuel derived from the combustible fraction of municipal solid waste.

**Resource Recovery.** The separation, extraction and reclamation of usable materials or energy from municipal waste.

**Resource Recovery Facility (Waste-to-Energy Facility).** A physical plant in which solid or liquid wastes are converted into usable materials, including steam energy, fuels, and raw commodities.

**Roll-Off Bin.** A refuse storage container which can be hydraulically pulled onto the bed of a vehicle for efficient handling and transport.

**Rubbish.** Nonputrescible waste, including paper, aluminum cans, wood, glass and like materials.

**Sanitary Landfill/Landfill/Fill.** As defined by the American Society of Civil Engineers: "A method of disposing of refuse on land without creating nuisances or hazards to public health or safety, by utilizing the principles of engineering to confine the refuse to the smallest practical volume, and to cover it with a layer of earth at the end of each day's operation, or at such more frequent intervals as may be necessary."

**Scrubber Sludge.** A smokestack's residual acidic gas and/or particulate solids (in the form of calcium sulfite, calcium sulfate and ash) which serve a chemical cleaning function.

**Sewage Sludge.** Chemically bound solid deposits extracted from the wastewater of sewage treatment plants.

**Shredding.** A process in which refuse is mechanically fragmented into 1-2 inch diameter pieces for the purpose of increasing waste compaction density. Compacted waste may not require landfill cover to deter disease vectors such as rats or gulls. Shredders include: hammermills, drum and wet pulverizers and crushers.

**Shredfill.** A shredded solid waste disposal site in which fill density may be increased by 25% due to compaction alone, such that daily cover may not be required. Shredfills often have longer functional lives.

**Solid Waste.** The material discards of residential, commercial and industrial sectors, including garbage, trash, demolition and construction refuse, appliances, automobiles and sewage sludge.

**Sorting.** Separation, into uniform categories, of secondary (reclaimable) commodities from solid waste for recycling.

**Source Reduction (waste reduction).** The decreased generation of waste, at the site of manufacture, following the introduction of new production technology, packaging or institutional design.

**Stoker.** A movable grate designed to transport a fuel bed through an incineration furnace.

**Sulfur Dioxide ($SO_2$).** A colorless, malodorous, toxic gas constituting a signficant air pollution threat under certain conditions.

**Thermocline.** That region, below the mixed layer in body of water, characterized by rapidly changing temperatures.

**Tipping.** The act of dumping garbage from sanitation trucks.

**TPD (Tons Per Day).** A unit expressing solid waste generation rates.

**Toxicity.** The quality of being poisonous. The United States Academy of Sciences defines the toxicity of a given material using the following parameters:

1. rate of release to the environment
2. residence time in the environment
3. potential for bioaccumulation
4. adverse effects on health

**Toxic Metals.** Metals (usually of the heavy metal class) which interefere with the respiration, metabolism or growth of organisms.

**Transfer Station.** A site, intermediate between waste collection and disposal, at which refuse is transferred from small to large capacity vehicles for further transport.

**Trash.** All nonputrescible disgards, including paper, wood, scrap metals, glass and similar savable materials.

**Tromelling.** The process of removing fine/dense material (glass fragments, grit, etc.) from shredded refuse. An open-ended screened drum (the tromel) allows small particles to pass as refuse is tumbled within.

**Urban Ore.** Ferrous scrap, mixed nonferrous metals and glass aggregate of market value, reclaimed from municipal solid waste.

**User Fee (tariff).** Surcharge imposed for services rendered.

**Vector.** An organism that can transmit pathogens to man.

**Waste Commodity.** Any material, reclaimed from waste, of actual or potential use.

**Waste Utilization.** Transformation of waste commodities into new products.

**Wastewater.** The aqueous effluent generated by resource recovery facilities, which generally requires chemical treatment before discharge to sanitary sewage systems.

**Wet Scrubber.** A pollution control device which removes particulate and gaseous impurities from flue gas using either a spray or a wetted impaction surface. (See Scrubber Sludge).

# REFERENCES

## CHAPTER 1


1. Franklin, M.A., N.S. Artz, P.E. Hunt and R.G. Hunt. 1986. Characterization of Municipal Solid Waste in the United States, 1960 to 2000. Franklin Associates, Ltd. Prairie Village, Kansas; Prepared for the U.S. Environmental Protection Agency.

2. Tchobanoglous, G., H. Theisen and R. Eliassen. 1977. Solid wastes: Engineering principles and management issues. McGraw Hill, New York. 621 pp.

3. U.S. Bureau of the Census. 1984. Statistical Abstract of the United States: 1985 (105th edition). Washington, D.C. 991 pp.

4. U.S. Bureau of the Census. 1984. op. cit.

5. Dogett, R.M., M.K. O'Farrell and A.L. Watson. 1980. Forecasts of the quantity and composition of solid waste. U.S. Environmental Protection Agency, Cincinatti (OH), EPA-600/5-80-001. 171 pp.

6. Chandler, W.V. 1983. Materials recycling: The virtue of necessity. Worldwatch Paper 56. Worldwatch Institute, Washington, D.C. 52 pp.

7. Chandler. 1983. Ibid.

8. White, P. T. 1983. The fascinating world of trash. Natl. Geogr. 163 (4):424-457.

9. G. Boyd, New York State Legislative Commission on Solid Waste Management, personal communication, 1985.

10. E. Santoro, U.S. Environmental Protection Agency, Region 2, 1986; M. Greges, U.S. Army Corps of Engineers, New York District, personal communication, 1986.

11. Smith, G., U.S. Environmental Protection Agency, private communication, 1985.

12. New York City Department of Sanitation. 1985. Final environmental impact statement for the proposed resource recovery facility at the Brooklyn Navy Yard.

13. Chanlett, E.T. 1979. Page 67 in Environmental Protection. (2nd edition). McGraw Hill, New York, 569 pp.

14. Chanlett. 1979. Ibid.

15. Chanlett. 1979. Ibid.

16. Skjei, E. and M. D. Whorton. 1983. Of mice and molecules: Technology and human survival. Dial Press, New York. 347 pp.

## CHAPTER 2

1. Melosi, M.V. 1981. Garbage in the cities: Refuse, reform and the environment, 1880-1980. Texas A&M University Press.

2. Gunnerson, C.G. 1973. Debris accumulation in ancient and modern cities. J. Environ. Eng. Div., ASCE, 99(EE3) 229-243.

3. Gunnerson, C.G. 1973. Ibid.

4. Gunnerson, C.G. 1973. Ibid.

5. Gunnerson, C.G. 1973. Ibid.

6. Melosi, M.V. 1981. op. cit.

7. Melosi, M.V. 198l. op. cit.

8. Melosi, M.V. 1981. op. cit.

9. Melosi, M.V. 1981. op. cit.

10. Melosi, M.V. 1981. op. cit.

11. "How hazardous are municipal wastes?" Amer. City and Country, March, 1983, 98:41-42.

12. Franklin, W.E., M.A. Franklin and R.G. Hunt. 1982. Waste Paper: The future of a resource, 1980-2000. Franklin Associates, Ltd. Prairie Village, Kansas for the Solid Waste Council of the Paper Industry. American Paper Institute.

13. Franklin, W.E. et al. 1982. Ibid.

14. Edwards, Victor H. "Potential useful products from cellulose material." Biotechnology and Bioengineering Symposium No. 5. pp. 321-338 in C.R. Wilke (ed.), Cellulose as a Chemical and Energy Resource. John Wiley (1975)

15. Epstein, S.S., L.O. Brown and C. Pope. 1982. Hazardous Waste in America. Sierra Club Books, San Francisco, 593 pp.

16. Expert Advisory Committee on Dioxins. 1983. Report of the Joint Health and Welfare Canada/Environment Canada Expert Advisory Committee on Dioxins.

17. Chanlett, Emil T. 1979. Page 2 in Environmental Protection, McGraw Hill Book Co., New York. 569 pp.

18. U.S. Bureau of the Census. 1980. Statistical Abstract of the United States: 1981 (102nd Edition) Washington, D.C.

19. U.S. Bureau of the Census. 1984. Statistical Abstract of the United States: 1985 (105th Edition) Washington, D.C. 991 pp.

20. Searl, Milton F.; Technical Manager, Energy Study Center, Electric Power Research Institute, Palo Alto, California. 1986, personal communication.

21. Franklin, W.E. et al. 1982. Ibid.

22. Hayes, D. 1978. Repairs, Reuse, Recycling--First Steps Toward a Sustainable Society. Worldwatch Paper 23. Worldwatch Institute. Washington, D.C. 45 p.

23. Food and Drug Administration. 1985. Pulp from reclaimed fiber. Code of Federal Regulations. Title 21, Part 176.260. (1 April 1985, edition.)

24. Hayes, D. 1978. op. cit.

25. Edwards, Victor H. "Potential useful products from cellulose material." 1975. op. cit.


## CHAPTER 3


1. Steisel, N.; Commissioner. New York City Department of Sanitation. 1985. personal communication.

2. Steisel, N. 1985. Ibid.

3. New York City Department of Sanitation. 1985. Final Environmental impact statement for the proposed resource recovery facility at the Brooklyn Navy Yard.

4. Steisel,N. 1985. op. cit.

5. U.S. Environmental Protection Agency. 1985. Control of air pollution from new motor vehicles and new motor vehicles engines; gaseous emission regulations for 1987 and later model year light duty vehicles, and for 1988 and later model year light-duty trucks and heavy-duty engines; particulate emission regulations for 1988 and later model year heavy-duty diesel engines. (15 March 1985). Fed. Reg. 50(51):10606-10708.

6. Bumb, R.R., W.B. Crummett, S.S. Cutie, J.R. Gledhill, R.H. Hummel, R.O. Kagel, L.L. Lamparski, E.V. Luoma, D.L. Miller, T.J. Nestrick, L.A. Shadoff, R.H. Stehl and J.S. Woods. 1980. Trace chemistries of fire: a source of chlorinated dioxins. Science. 210(4468):385-390.

7. Commoner, B., M. McNamara, K. Shapiro and T. Webster. 1984. Environmental and ecomonic analysis of alternative municipal solid waste disposal technologies, IV. The risks due to emissions of chlorinated dioxins and dibenzofurans from proposed New York City incinerators. Center for the Biology of Natural Systems, Queens College, City University of New York. 28 pp.

8. Public Works. 1984. Newspaper recycling system saves county landfill space. Public Works 115(3):77.

9. Millar, A. 1982. Residential solid waste collection. Urban Data Service Reports, International City Management Association, Washington, D.C. 14(12): 9 pp.

## CHAPTER 4

1. Chanlett, E.T. 1979. Environmental Protection, 2nd ed., McGraw Hill, New York. 569 pp.

2. Hall, C.W. 1984. Groundwater quality protection: the issue in perspective. Environ. Prof. 6:46-51.

3. Skjei, E. and M.D. Whorton. 1983. Pages 126-127 in Of mice and molecules: technology and human survival. Dial Press, New York. 347 334 pp.

4. Skjei, E., and M.D. Whorton. 1983. Ibid.

5. Chanlett, E.T. 1979. Environmental Protection, 2nd ed., McGraw Hill, New York. 569 pp.

6. Chanlett, E.T. 1979. Ibid.

7. Chanlett, E.T. 1979. Ibid. Pages 81-82.

8. Epstein, S.S., L.O. Brown and C. Pope. 1982. Hazardous Waste in America. Sierra Club Books, San Francisco, 593 pp.

9. Hall, C.W. 1984. op. cit.

10. Hall, C.W. 1984. op. cit.

11. Hall, C.W. 1984. op. cit.

12. Bagchi, A. 1983. Design of natural attenuation landfills. J. Environ. Eng. Div., ASCE, 109(4):800-811.

13. Bagchi, A. 1983. Ibid.

14. Brunner, C.R. 1985. Hazardous air emissions from incineration. Chapman and Hall, New York. 222 pp.

15. Brunner, C.R. 1985. Ibid.

16. Brunner, C.R. 1985. Ibid.

17. Brunner, C.R. 1984. Ibid. p. 66-76.

18. Chanlett, E.T. 1979. op. cit. p. 356.

19. Burger, J. 1982. Report on bird control at J.F. Kennedy International Airport. Unpublished report prepared for the Port Authority of New York and New Jersey. Rutgers University, New Brunswick, New Jersey. 113 pp.

20. Burger, J. 1982. Ibid.

21. Van Tets, G. F. 1969. Quantitative and qualatative changes in habitat and avifauna at Sydney Airport. CSIRO Wildl. Res., v. 14, 117-28

22. Malione, B.R. 1985. Environmental impact of sanitary landfills. Working Paper 17. Marine Sciences Research Center, State University of New York at Stony Brook.

23. Light, L. 1985. Pumping methane dollars from dumps. Newsday. 10 July 1985, p. 88.

References

24. ASTM, Subcommittee D18.14 on Soil and Rock Pollution. 1918. Hydrogeological view of waste disposal in the shallow subsurface. Geotech. Test. J. 4(2):53-57.

25. Malione, B. 1985. op. cit.

26. Malione, B. 1985. op. cit.

27. Diaz, Luis, G.M. Savage, and C. G. Golueke. 1982. Resource Recovery from Municipal Solid Wastes. Volume II: Final Processing. CRC Press, Inc. 178 pp.

28. Hagerty, D.J., J.L. Pavoni, and J.E, Heer, Jr. 1973. Solid waste management. Van Nostrand Reinhold Co., New York. 302 pp.

**CHAPTER 5**

1. Slader, W.J.L., C.M. Menzie and W.I. Reichel. 1966. DDT residues in adelie penguins and crabeater seal from Antarctica: Ecological implications. Nature 210(5037):670-673.

2. Park, P.K, and T.P. O'Connor. 1981. Ocean dumping research: Historical and international development. Pages 3-23 <u>in</u> B.H. Ketchum, D.R. Kester and P.K. Park (eds.), Ocean Dumping of Industrial Wastes. Proceedings of the First Ocean Dumping Symposium, October 10-13, 1978, University of Rhode Island, West Greenwich, Rhode Island. Plenum Press, New York. 555 pp.

3. Les Barthlow, Maritime Labor Data Analyst. U.S. Dept. of Labor and Training. personal communication, 1986.

4. U.S. Dept. of Transportation. 1985. Boating statistics 1984. U.S. Coast Guard. COMDTINST M16754.1F Washington, D.C. 33 pp.

5. Matthews, W. 1975. Marine litter. Pages 405-438 <u>in</u> A report of the study panel on assessing potential ocean pollutants to the Ocean Affairs Board, Commission on Natural Resources. Natl. Res. Coun., Natl. Acad. Sci., Washington, D.C.

6. Carpenter, E.J. and K.L. Smith. 1972. Plastics on the Sargasso Sea. Science 175(4027):1240-1241.

7. Colton, J.B., Jr. 1974. Plastics in the oceans. Oceanus 18(1):61-64.

8. Venrick, E.L., T.W. Backman, W.C. Bartram, C.J. Platt, M.S. Thronhill and R.E. Yates. 1973. Man-made objects on the surface of the north central Pacific Ocean. Nature 241(5387):271.

9. Wong, D.C., D.R. Green and W.J. Cartney. 1974. Quantitative tar and plastic waste distributions in the Pacific Ocean. Nature 247(5435):30-32.

10. Scott, G. 1975. The growth of plastic packaging litter. Int. J. Environ. Stud. 7(2):131-132.

11. Dixon, T.R. and A.J. Cooke. 1977. Discarded containers on a Kent beach. Mar. Pollut. Bull. 8(5):105-109.

12. Gregory, M.R. 1977. Plastic pellets on New Zealand beaches. Mar. Pollut. Bull. 8(4):82-84.

13. Hays, H. and G. Cormans. 1973. Plastic particles found in tern pellets on coastal beaches and at factory sites. Linn. Newslett. 27:44-46.

14. Dixon, T.R. 1978. Shoreline refuse. Mar. Pollut. Bull. 9(6):145.

15. Cundell, A.M. 1973. Plastic material accumulating in Narragansett Bay. Mar. Pollut. Bull. 4(12):187-188.

16. Hays, H. and G. Cormans. 1973. op. cit.

17. Rothstein, S.I. 1973. Plastic particle pollution of the Atlantic Ocean: Evidence from a seabird. Condor 75(1):344-345.

18. Parslow, J.L.F., D.J. Jeffries and M.C. French. 1972a. Elastic thread pollution of puffins. Mar. Pollut. Bull. 3(3):43.

19. Parslow, J.L.F., D.J. Jeffries and M.C. French. 1972b. Injected pollutants in puffins and their eggs. Bird Study 19:18-33.

20. Carpenter, E.J., S.L. Anderson, G.R. Harvey, H.P. Miklas and B.B. Peck. 1972. Polystyrene spherules in coastal waters. Science 178(4062):749-750.

21. Karter, S., R.A. Milne and M. Bainsberry. 1973. Polystyrene waste in the Severn Estuary. Mar. Pollut. Bull. 4(9):144.

22. Karter, S., F. Abou-Seedo and M. Sainsbury. 1976. Polystyrene spherules in the Severn Estuary--a progress report. Mar. Pollut. Bull. 7:52.

23. Colton, J.B., Jr. 1974. op. cit.

24. Colton, J.B., Jr., F.D. Knapp and B.R. Burns. 1974. Plastic particles in surface waters of the northwest Atlantic. Science 185:491-497.

25. Waldichuk, M. 1977. Plastic and seals. Mar. Pollut. Bull. 9:197.

26. Kottcamp, G.M. and P.B. Moyle. 1972. Use of disposable beverage cans by fishermen in the San Joaquin Valley. Trans. Am. Fish. Soc. 101(3):566.

27. Brasher, J., Jr. 1973. Offshore industry spurs sportfishing. Offshore 33:122.

28. Salazar, . 1973. Animal attraction to sunken submarines. Oceans 6(3):68-70.

29. Texas A&M University. 1973. Avoiding snags saves $$. The University and the Sea 6:5.

30. Carpenter, E.J. 1978. Persistant solid synthetic materials, in particular, plastics, which may interfere with any legitimate use of the sea. Pages 55-74 in Data profiles for chemicals for the evaluation of the environment of the Mediterrarean Sea. Volume II. International Register of Potentially Toxic Chemicals (IRPTC). United Nations Environment Programme, Geneva, Switzerland. 108 pp.

31. H.M. Stanford, Oceans Assessment Division, National Oceanic and Atmospheric Administration, personal communication, 1985.

32. Swanson, R.L., H.M. Stanford, J.S. O'Connor, S. Chanesman, C.A. Parker, P.A. Eisen and G.F. Mayer. 1978. Pollution of Long Island ocean beaches. J. Environ. Eng. Div., ASCE, 104(EE6):1067-1085.

33. Jewett, S.C. 1976. Pollutants of the Northeast Gulf of Alaska. Mar. Pollut. Bull. 7(9):169.

34. Feder, H.M., S.C. Jewett and J.R. Hilsinger. 1978. Man-made debris on the Bering Sea floor. Mar. Pollut. Bull. 9(2):52-53.

35. Van Banning, P. 1972. The continental shelf: a future refuse bin? Vadbl. Biol. 19:392-394.

36. Lawson, E. and R. Whitesides, Jr. 1972. Briefing document for the President's water pollution control advisory board. EPA Region 2.

37. Lawson, E. and R. Whitesides, Jr. 1972. op. cit.

38. Lawson, E. and R. Whitesides, Jr. 1972. op. cit.

39. IEC - Oceanics 1973. Ocean disposal in select geographic areas. Contract 68-01-0769. IEC Report 4460 c 1541.

40. First, M.W. 1972. Municipal waste disposal by shipborn incineration and sea disposal of residues. Department of Environmental Health

Sciences. Harvard University School of Public Health.

41. Pratt, S.D., S.B. Saila, A.G. Gaines, Jr. and J.E. Krout. 1973. Biological effects of ocean disposal of solid wastes. Marine Technical Report. University of Rhode Island, Kingston, Rhode Island. 53 pp.

42. Rowe, G.T., Brookhaven National Laboratory, personal communication, 1985.

43. Pratt, S.D. et al. 1973. op. cit.

44. Saker, M., Jr. 1978. Economic evaluation of ocean reef building and option for combusion waste disposal. Prepared for the CWARP. unpublished.

45. Devanney, J.W., III, V. Livanos and J. Pattell. 1970. Economic aspects of solid waste disposal at sea. Sea Grant Report. #MITSG 71-2. Index #71-602-new. Massachusetts Institute of Technology, Cambridge, Massachusetts.

46. Metcalf and Eddy, Inc. 1969. Report to the New York State Pure Water Authority on nail haul disposal of solid waste for Westchester County, New York. February 27.

47. First, M.W. 1972. op. cit.

48. Devanney, J.W. III et al. 1970. op. cit.

49. Loder, T.C., F.E. Anderson and T.C. Shevenell. 1973. Sea monitoring of emplaced baled solid waste. University of New Hampshire Report, S.D.-118. 107 pp.

50. Loder, T.C. et al. 1983. Ibid.

51. Loder, T.C. et al. 1983. Ibid.

52. Pratt, S.D. et al. 1973. op. cit.

53. Pratt, S.D. et al. 1973. op. cit.

54. Pratt, S.D. et al. 1973. op. cit.

55. First, M.W. 1969. Waste incineration at sea and ocean disposal of nonfloating residues. Department of Environmental Health Sciences. Harvard University School of Public Health.

56. Oviatt, C.A. 1968. The effects of incinerator residue on selected marine species. Pages 108-100 in Proceedings of the Annual N.E. Region Antipollution Conference. July 22-24, 1968. University of Rhode Island,

Kingston, Rhode Island.

57. First, M.W. 1969. op. cit.

58. Bamber, R.N. 1980. A summary of the effects of dumped pulverized fuel ash on the benthic fauna of the Northumberland Coast. Central Electricity Generating Board (CERL) Job #JV440 MBL Fawley Report #R.D./L/R 2019. 19 pp.

59. Bamber, R.N. 1980. Ibid.

60. Woodhead, P.M.J. and F.J. Roethel. 1985. Characterization and stabilization of Odgen - Chicago residue. Working Paper 14, Marine Sciences Research Center, State University of New York at Stony Brook, New York. 33 pp.

61. Woodhead, P.M.J., J.H. Parker, H.P. Carlton and I.W. Duedall. 1984. Coal-waste artificial reef program, phase 4B. Electric Power Research Institute, EPRI CS-3726, Project 1341-1, Interim Report (November 1984). 272 pp.

62. Woodhead, P.M.J. et al. 1984. Ibid.

63. Bogost, M.W. 1971. Hawaii's experiment with ocean disposal of baled waste. Mar. Technol. Soc. 7(1):34-37.

64. Bogost, M.W. 1971. op. cit.

**CHAPTER 6**

1. New York City Department of Sanitation. 1985. Page 1.1 in Final environmental impact statement for the proposed resource recovery facility at the Brooklyn Navy Yard.

2. New York City Department of Sanitation. 1985. Page 1.23 in Final environmental impact statement for the proposed resource recovery facility at the Brooklyn Navy Yard.

3. Licata, A. 1986. "Designing for good combustion", 24 January, 1986 Municipal Solid Waste Forum, Marine Sciences Research Center, State University of New York at Stony Brook.

4. Licata, A. 1986. Ibid.

5. Skjei, E., and M. D. Whorton. 1983. Page 95 in Of mice and molecules: Technology and human survival. Dial Press, New York. pg. 95

6. Skjei, E., and M. D. Whorton. 1983. Ibid.

7. New York City Department of Sanitation. 1985. Page 2.38 in Final environmental impact statement for the proposed resource recovery facility at the Brooklyn Navy Yard.

8. New York City Department of Sanitation. 1985. Ibid.

9. Skjei, E., and M. D. Whorton. 1983. Page 99 in Of mice and molecules: Technology and human survival. Dial Press, New York.

10. New York City Department of Sanitation. 1985. Page 2.38 in Final environmental impact statement for the proposed resource recovery facility at the Brooklyn Navy Yard.

11. Skjei, E., and M. D. Whorton. 1983. Page 101 in Of mice and molecules: Technology and human survival. Dial Press, New York.

12. New York City Department of Sanitation. 1985. Page 2.33 in Final environmental impact statement for the proposed resource recovery facility at the Brooklyn Navy Yard.

13. Calculated from data provided by Christopher Gross of the Long Island Lighting Company.

14. Greenburg, R. R., W. H. Zeller and G. E. Gordon. 1978. Composition and size distributions of particles released in refuse incineration. Environ. Sci. Technol. 12(5):566-573.

15. Cambell, W. J. 1976. Metals in the waste we burn. Environ. Sci. Technol. 10(5):436-439.

16. Law, S. L. and G. E. Gordon. 1979. Sources of metals in municipal incinerator emissions. Environ. Sci. Technol. 13(4):432-438.

17. Richard, J. J. and G. A. Junk. 1981. Polychlorinated biphenyls in effluents from combustion of coal/refuse. Environ. Sci. Technol. 15(9):1095-1100.

18. Richard, J. J. and G. A. Junk. 1981. Ibid.

19. Brunner, C. R. 1985. Page 58 in Hazardous air emissions from incineration. Chapman and Hall, New York.

20. Brunner, C R. 1985. Page 59 in Hazardous air emissions from incineration. Chapman and Hall, New York.

21. Brunner, C. R. 1985. Page 62 in Hazardous air emissions from in-

cineration. Chapman and Hall, New York.

22. Benfenati, E., F. Gizzi, R. Reginato, M. Lodi and R. Tagliaferri. 1983. PCCDs and PCDFs in emissions from an urban incinerator: correlation between concentration of micropollutants and combustion conditions. Chemosphere. 12(9):1151-1157

23. Penner S. S., and D. Wiesenhahn. 1985. Local and global implications of production of dioxins and furans by properly operating Martin-type municipal waste incinerators. University of California at San Diego. La Jolla, CA. 12pp.

24. Expert Advisory Committee on Dioxins. 1983. Report of the Joint Health and Welfare Canada/Environment Canada Expert Advisory Committee on Dioxins.

25. New York City Department of Sanitation. 1985. Page 2.46 in Final environmental impact statement for the proposed resource recovery facility at the Brooklyn Navy Yard.

26. Fred C. Hart Associates, Inc. Assessment of potential health impacts associated with predicted emissions of polychlorinated dibenzo-dioxins and polychlorinated dibenzo-furans from the Brooklyn Navy Yard Resources Recovery Facility. New York.

27. New York Academy of Sciences. 1984. Proceedings of the Resource Recovery Policy Dialogue. (18 December 1984.) New York.

28. Tchobanoglous, G., H. Theisen, and R. Eliassen. 1977. Page 300 in Solid wastes: Engineering principles and management issues. McGraw Hill, New York. 321pp.

29. Roethel, F. J., I.W. Duedall and P. M. J. Woodhead. 1983. Coal waste artificial reef program: Conscience Bay studies. Electric Power Research Institute. Palo Alto, CA.

30. Vence, Thomas D. 1984. Potential of recycling ash from resource recovery facilities in California, in Proceedings of the 1984 National Waste Processing Conference, American Society of Mechanical Engineers.

31. Tchobanoglous, G., H. Theisen, and R. Eliassen. 1977. Page 290 in Solid wastes: engineering principles and management issues. McGraw Hill, New York. 321pp.

32. Cameron, R.D. and F.A. Koch. 1980. Toxicity of landfill leachate. J. Water Poll. Con. Fed. 52(4):760-764.

33. Cameron, R.D. and F.A. Koch. 1980. Ibid.

## CHAPTER 7

1. National Solid Waste Management Association 1986. Tipping Fee Survey. Waste Age 17(3); 58-60.

## CHAPTER 8

1. Chandler, W.U. 1983. Materials recycling: the virtue of necessity. Worldwatch Paper 56. October 1983. Worldwatch Institute. Washington, D.C. 52 pp.

2. Tapscott, G. 1984 States should plan to reduce wastes. Manage. World Wast. 27(6):44.

3. Tapscott, G. 1984 States begin stressing waste reduction programs. Manage. World Wast. 27(2):34-35.

4. Doggett, R.M., M.K. O'Farrell, and A.L. Watson 1980. Forecasts of the quality and composition of solid waste. U.S. EPA, Cincinatti. EPA-600/5-80-001. 171 pp.

5. Chandler. 1983. op. cit.

6. Chandler. 1983. op. cit.

7. Conn, W.D. and E.C. Warren. 1979. Developing a tentative model of disposal decisions. J. Environ. Sys. 9(2):129-144.

8. Henstock, M.E. 1980. Some barriers to the use of materials recovered from municipal solid waste. Resour. Pol. 6(3):240-252.

9. Henstock, M.E. 1980. op. cit.

10. Jacobs, H.E. and J.S. Bailey. 1982 Evaluating participation in a residential recycling program. J. Environ. Sys. 12(2):141-152.

11. Nesheim, E.E. and H.M. Theisen. 1983. Careful planning required for recovery projects. Manage. World Wast. 26(10):26-29.

12. Albrecht, O.W., E.H. Manuel, and F.W. Efaw. 1981. Recycling in the USA: vision and reality. Resour. Pol. 7(3):188-196.

13. Roth, L. 1983. Transfer station success for city of 2,662. Manage. World Wast. 26(4):102-105.

References


14. Bracken, B.D., N.A. Speed, and R. Young. 1981. Conceptually designing a solid waste transfer station. Public Works 112(5):60-64.

15. Blanker, W.P. 1983 Refuse volume reduction through mandatory recycling. Public Works 14(8):57-59.

16. Grogan, P. 1983 Successful community recycling: Is it Possible? Public Works 114(1):50-52.

17. Larkey, B. 1984 Solid waste management and recycling on a countywide basis. Public Works 115(7):52-64.

18. Tchobanoglous, G., H. Theisen, and R. Eliassen. 1977. Solid wastes: Engineering principles and management issues. McGraw Hill, New York. 621 pp.

19. Larkey. 1984. op. cit.

20. Larkey. 1984. op. cit.

21. Public Works. 1984. Newspaper recycling system saves county landfill space. Public Works 115(3):77.

22. Public Works. 1984. op. cit.

23. Gill, G. and K. Lahiri. 1980. An econometric model of wastepaper recycling in the USA. Resour. Pol. 6(4):320-325.

24. Grogan. 1983. Ibid.

25. Hayes, D. 1978. Repairs, Reuse, Recycling--First Steps Toward a Sustainable Society. Worldwatch Paper 23. September 1078. Worldwatch Institute, Washington, D.C. 45 pp.

# INDEX